MASSEY-FERGUSON
Tractors

MASSEY-FERGUSON
Tractors

MICHAEL WILLIAMS

Special Photography by David Williams

FARMING PRESS

First published in the U.K. 1987 by Blandford Press
Copyright © 1987 Michael Williams

Reprinted 1989, 1990, 1992 by Farming Press

British Library Cataloguing in Publication Data

Williams, Michael, *1935 Nov. 4-*
  1. Massey-Ferguson tractors—History
  I. Title
  629.2'25   TI233.5

ISBN 0 85236 203 X

Published by Farming Press Books
Wharfedale Road, Ipswich IP1 4LG

Distributed in North America
by Diamond Farm Enterprises,
Box 537, Alexandria Bay, NY 13607, USA

Printed and bound in Great Britain by
Butler and Tanner Ltd, Frome and London

CONTENTS

ACKNOWLEDGEMENTS

Much of the material for this book was made available by Massey-Ferguson, and I would particularly like to thank John Briscoe at Stoneleigh and Pedr Bjerre at the Varity Corporation offices in Toronto for helping me to find information and pictures I needed.

Thanks also to Peter Ledwith, Curator at the Ontario Agricultural Museum, who helped me find my way around the superb collection of Massey-Harris tractors and machinery at the Museum, and allowed me to use the archives there.

The title of the last chapter in this book, A Global Corporation, is borrowed with thanks from the book of the same name by Prof. E.P. Neufeld, published by the University of Toronto Press.

Most of the photographs used in this book are from the company archives, but the special photography was by David Williams, illustrating tractors at the Massey-Ferguson Museum at Stoneleigh, the exceptionally original Ferguson Type A belonging to A.T. Oliver & Sons at their Shefford, Bedfordshire, premises, and beautifully restored tractors owned by Tim Gooding of Needham Market, Suffolk, and Ben Sergeant of Peterborough.   MW

ILLUSTRATIONS
The following sources are acknowledged with thanks.

Colour Plates
Massey-Ferguson: Nos 7, 11, 24, 25, 27, 28, 30, 31, 32 and 33
Andrew Morland: Nos 3, 4, 10 and 13
David A. Williams: Nos 8, 17, 18, 19, 20, 21, 22, 23 and 26
Michael R. Williams Collection: Nos 1, 2, 5, 6, 9, 12, 14, 15 and 16

Black and White
Crown Copyright: page 63 (upper).
Institute of Agricultural History and Museum of English Rural Life: pages 16, 27, 53 (lower), 61 and 65 (top).
Landini: pages 108 and 109.
David A. Williams: pages 2, 35, 37, 50 (upper), 64, 65 (lower), 66, 67, 68, 72, 73, 77 (upper), 79 (upper), 81, 85, 87 (lower), 88, 94, 103, 104.
Michael R. Williams Collection: pages 12, 17, 20, 22 (lower), 28, 45, 47 and 69.
Andrew Morland: pages 53 (upper) and 75.
Perkins Engines Ltd: pages 101, 102, 114 (lower), 115 and 116.
Massey-Ferguson Archives: all other photographs.

INTRODUCTION

Few companies in the tractor industry have a longer history than Massey-Ferguson, and there are none which make a bigger contribution to power farming in the western world.

The growth, from two small workshops supplying simple implements for Canada's pioneer farmers to the world's biggest tractor company, has involved some of the farm equipment industry's most talented and successful people.

This book is about some of those people and the machines and tractors they built. The story starts with the Massey and Harris families, whose enterprise took the company outside Canada to become one of the world's best-known farm machinery manufacturers. Success in the tractor market came later, when Massey-Harris acquired the rights to the Wallis tractors with the distinctive U-frame which had been one of the most significant new developments in tractor design.

Meanwhile, in Italy, the Landini family was building some of Europe's biggest tractors, Frank Perkins in England was helping to pioneer the benefits of diesel power, and Harry Ferguson was developing the ideas which have revolutionised tractor efficiency through his Ferguson System of implement attachment and control.

These developments are all part of the history of farm tractor development and have made their contributions to the Massey-Ferguson success story.

1 · THE PIONEERS

The Massey-Ferguson organisation is the western world's biggest tractor manufacturer but the company's history started when horses, mules and oxen were the only power on the land and the first farm tractors had not been developed.

Mergers and takeovers have played an important part in the growth of the company and the most significant of these was in 1891, when the Massey Manufacturing Company of Toronto and A. Harris, Son and Company of Brantford merged to form the Massey-Harris Company, easily the largest farm machinery manufacturer in Canada.

Both the Massey and the Harris companies had been among the pioneers of the farm equipment industry in Canada. Daniel Massey had started his business in about 1847 when he bought a workshop and equipment to do repairs and to make simple implements to sell to local farmers. Alanson Harris bought a small foundry and began making farm machinery in 1857.

There were some significant similarities in the way the Massey and the Harris companies developed. Both had started in a small way, but both expanded rapidly and soon had to move to bigger premises so that production could be increased to keep pace with demand. Most of this rapid growth came from harvesting equipment and the Massey and Harris companies became the market leaders in Canada for equipment such as mowers, hay rakes, reapers and binders.

The early emphasis on harvesting machinery continued to be important long after the two companies had joined forces and it led directly to the success of the Massey-Harris reaper-threshers in the 1920s and the world's first self-propelled combine, the MH 20, in 1938.

Another feature of the early development of the Massey and Harris companies was their interest in finding overseas markets. This started in 1867 when Massey harvesting machines were chosen to represent Canada at the great International Exposition in Paris. The event was a triumph for the Massey company which won two Grand Gold medals, one for its static display of machines and the second for its performance in a working demonstration.

It was the first time Massey equipment had been in direct competition with leading American and European machines and the results showed that the Canadian company's products were among the best in the world. Other international shows followed, including the Indian and Colonial Exhibition of 1886 in London, where both Massey and Harris machines were displayed and brought additional export orders.

Further evidence of the quality of Canadian farm machinery came in

1889 when a Massey binder took part in a 4-day harvesting trial organised by the French Ministry of Agriculture. The event, claimed to be the biggest field trial of its kind, was part of the programme for the Paris Universal Exposition and attracted entries from all the leading manufacturers. The Massey Toronto Light Binder was the outright winner and the company was awarded a special trophy and a gold medal, and gained valuable publicity and prestige.

The Massey and Harris companies were the first farm equipment manufacturers in Canada to develop substantial overseas markets, and this policy was continued after the merger to produce a company with a completely international approach to production and marketing.

The 1891 merger brought together the two biggest farm machinery companies in Canada and it was soon followed by a succession of take-overs which extended the Massey-Harris product line into ploughs, discs, wagons and manure-spreaders.

By the turn of the century Massey-Harris had become one of the leading companies in the industry, with an international reputation for harvesting equipment, but there was still no attempt to move into the power farming market. Agricultural steam engines had become important for driving stationary equipment, including threshing machines, and for ploughing, and tractors were being built in small numbers, but the Massey-Harris company remained outside this market which was dominated by American and British manufacturers.

There was, however, a family interest in steam engine production. The

The Massey Harvester was first manufactured in 1878.

Sawyer-Massey traction engine at the Ontario Agricultural Museum, Canada.

Massey family, which remained the biggest shareholders in the new Massey-Harris company, bought a 40 per cent share in the L.D. Sawyer Co. in 1892. The Sawyer company was based at Hamilton, Ontario, and started building portable steam engines in the 1860s. When the Massey family became involved the name was changed to the Sawyer-Massey Co. and production was expanded to make the company Canada's biggest and most successful manufacturer of agricultural steam engines and threshing equipment.

Although there appear to have been no formal links between the two companies there was certainly co-operation over sales promotion and marketing. Sawyer-Massey probably gained substantially from the association, which came to an abrupt end in 1910 when the Massey family withdrew their finanical interest in the Sawyer company.

The exact details of the Massey family involvement in Sawyer-Massey are obscure and it is not clear what caused the break in 1910. It has been suggested that the Masseys objected to a company decision to double the production of steam engines, but 1910 was also the year in which Sawyer-Massey started to build gasoline tractors and this may have been the factor which caused the rift.

Sawyer-Massey moved into the tractor market at just the right time. Demand was growing rapidly, especially during the war years when farmers were encouraged to mechanise in order to produce more food. The company built a range of sturdy tractors for the Canadian market until the early 1920s, when the flood of low-priced imports from the United States forced a change of policy and tractor production came to an end.

Meanwhile Massey-Harris was making its first move into power farming. It happened in 1910 when the Deyo-Macey gasoline-engine company of Binghamton, New York, was purchased. This provided a range of small hopper-cooled engines at a time when this type of power unit was selling in large numbers. The engines were made at Binghampton for several years and sold in Canada under the Massey-Harris name, but in 1916 production was transferred to a new Massey-Harris factory at Weston, Toronto, and the old Deyo-Macey property was sold. A few years later Weston became the first Massey-Harris tractor factory.

2 · BULL AND PARRETT

The Massey-Harris company's first venture into the tractor market came in 1917 with an agreement to import the Bull tractor from the United States to add to the Massey-Harris product line for the Canadian market.

In 1917 the demand for tractors in Canada was expanding rapidly and there must have been plenty of American manufacturers who would have welcomed Massey-Harris as their Canadian distributor. The Bull Tractor Company of Minneapolis had helped to pioneer the idea of a small, low-cost tractor and this had placed the company among the market leaders in the United States with a success story which must have impressed the Massey-Harris management.

The team behind the Bull company had previously started the Gas Traction Co. in Minneapolis which claimed to be the first company to design and manufacture a tractor powered by a four-cylinder engine. This was the Big 4, one of the most successful of the heavyweight tractors which helped to break up large areas of prairie land in the United States and Canada. Emerson Brantingham bought the Gas Traction Co. in 1912.

D. M. Hartsough, who had produced the original version of the Big 4 in 1904, designed the first of the Bull company's tractors in 1913. This was a small tractor powered by a horizontally-opposed two-cylinder engine producing 12 hp on the belt pulley and 5 hp at the drawbar. Mr Hartsough chose a three-wheeled design, with a large driving wheel, a small wheel at the front for steering and a medium-sized wheel at the rear.

The tractor was called the Little Bull, and it arrived on the American market in 1914 with a fanfare of publicity. The idea that there was a big potential demand for a small, low-cost tractor proved to be correct and the Little Bull was quickly among the best-selling models.

With the sales success came problems. The tractor, with its unconventional wheel arrangement, had been designed for ploughing, but the 5 hp available at the drawbar was not enough to achieve a worthwhile work rate and there were some design faults which earned the Little Bull a poor reputation for reliability.

In 1915 the company announced a new, more powerful model which was called the Big Bull, and this was the tractor which became part of the Massey-Harris product line in Canada 2 years later.

The Big Bull was simply a bigger, heavier version of the previous model. The engine design was still an opposed twin, but with the output raised to 25 hp, with 10 hp available at the drawbar. The tricycle arrangement of unequal-sized wheels, designed to keep the tractor level while ploughing, was retained with a 5 ft diameter driving wheel, a 2 ft 8 in wheel for steering and a 3 ft 4 in wheel to run on the unploughed ground.

The design included a cranked axle with an adjustment so that the level of the tractor could be altered for jobs other than ploughing.

With its extra power and an increase in weight to 4500 lb instead of the 3280 lb of the Little Bull, the new tractor appears to have been a success which the company proclaimed with much enthusiasm. It was advertised as 'The Bull With the Pull' and was described as 'the very last word in tractor construction'.

'If a veritable mountain of testimony or a veritable army of successful users, if an unequalled record of accomplishment in the field or if a record never yet attained by a tractor manufacturer will lend its influence in your consideration of the purchase of a light tractor, then we submit that we have more than 5000 testimonials in our file; we have between 9000 and 10 000 Bull Tractors now in use and more acres were plowed with the Bull tractor in the last two years than with any other machine in the world. These are facts well established and undisputed. If popularity counts for anything, you will decide in favor of The Bull With The Pull', the company declared in a sales leaflet published for the United States market in about 1917.

A special feature was the design of the pyramid-shaped steel points

Bull tractors working in England in about 1917.

attached to the rim of the big driving wheel. These helped the wheel to grip as it ran in the furrow bottom and they also helped to break up the soil below plough depth to reduce smear and compaction. This was a patented feature which, the company claimed, 'is the most sensational tractor improvement ever devised and is the most sensational feature of construction in the implement world today, placing the Bull at the head of deep-tilling machines'.

In Britain, where the tractor was called the Whiting–Bull, the publicity experts concentrated on the self-steering effect of the front wheel as it ran in the furrow bottom.

'The Whiting-Bull requires one attendant only – male or female – as the whole outfit (tractor and plough) is self-steering; a practical lesson in labour economy.'

By 1915 standards the Big Bull was probably a good tractor, and it might have been a success in Canada with the strength of the Massey-Harris marketing organisation behind it. In fact the sales agreement ended disastrously and it appears that very few Bull tractors were imported by Massey-Harris.

There were two main reasons for the failure. One of these was the fact that the Bull Tractor Company had subcontracted production to an outside supplier and this arrangement proved to be unreliable and production stopped in 1917, just when deliveries to Canada should have been increasing. After some delay, the Bull company was able to find an alternative supplier, but it was too late to save the Massey-Harris agreement.

Another reason for the failure was the impact of the new Fordson tractor, which was arriving in Canada in large numbers by the middle of 1918 and was making tractors like the Bull appear outdated and overpriced.

The Bull Tractor Co. never recovered fully from the financial effects of the supply disruptions and the company was out of business by about 1920. Meanwhile Massey-Harris had started another attempt to move into the tractor market, this time through an agreement with the Parrett Tractor Co. of Chicago, Illinois.

Under this agreement Massey-Harris began manufacturing tractors at their new engine factory in Weston on the outskirts of Toronto. The tractors were based on the design of brothers Dent and Henry Parrett who made their start in the tractor industry in 1913.

Parrett tractors had become well known in the United States with their distinctively-large front wheels. These look rather flimsy, but the Parrett company claimed the design had several advantages. The big circumference meant a larger surface area on the soil to reduce compaction, and it also meant less power was needed when climbing over obstacles. There was also reduced bearing wear, partly because the hubs were further away from the mud and dust, and because the extra diameter allowed a slower rotation speed.

The agreement between Massey-Harris and the Parrett company started in 1918 and production at the Weston plant began in 1919. The tractors were marketed in Canada under the Massey-Harris name, and there were three models which were called the MH 1, 2 and 3.

The MH 1 was based on the Parrett 12–25, a model which the American company had exported to Britain where it was sold in small numbers as the Clydesdale tractor. The design was based on a 7 in steel frame with a four-cylinder engine of $4\frac{1}{4}$ in bore and $5\frac{1}{2}$ in stroke with what the Massey-Harris leaflet described as a 'self-contained positive oiling system'.

Rear wheels were 5 ft in diameter with a 10 in wide rim, and the front

Illustrations from a 1917 Bull tractor
sales leaflet.

Clydesdale AGRICULTURAL TRACTORS

supersede the horse for all Farming Operations.

"The old order changeth giving place to the new."

Specifications, giving full particulars,
———can be obtained from : ———

R. MARTENS & Co., Ltd.
(MOTOR DEPT.),
Clydesdale House, 15a, Wilton Street,
Grosvenor Place, LONDON, S.W.

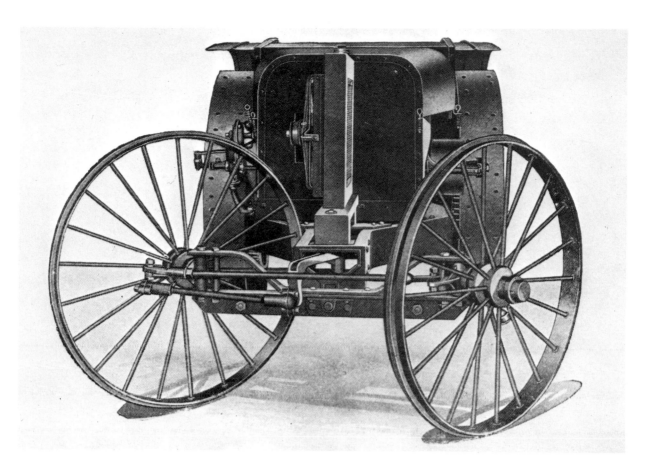

A sales leaflet illustration showing
the steering angle of the Parrett 12–
25.

Another publicity picture of the
Parrett 12–25.

Left:
Third and last of the Parrett tractor series from the Weston factory, Toronto.

Below left:
Driver's platform on the MH 1 tractor.

The MH 1, based on the Parrett 12–25.

wheel diameter was almost 4 ft with a 4 in rim. The two forward gears allowed travel speeds of 2.4 and 4 mph, with 1.8 mph in reverse.

When the tractor was used for ploughing, the front wheel running in the furrow bottom achieved a self-steering effect. This, said the Massey-Harris sales leaflet published in a less safety-conscious age, meant that the driver could dismount while the tractor was working in order to inspect the action of the ploughs or simply to keep warm in cold weather.

The MH 2 tractor was similar to the MH 1 model but with some fairly minor modifications. Both were powered by Buda four-cylinder engines, but the claimed power output was reduced from 12–25 to 12–22 hp in the later version. The high front wheels were retained, together with the sideways-facing radiator which gave the tractor its narrow front end to allow the wheels to steer. The lubrication system for the rear axle was improved and the gear ratios were changed to reduce the maximum forward speeds to 1.75 and 2.4 mph.

More substantial changes were made when the MH 3 model was introduced. The most obvious difference was repositioning the radiator to the more conventional transverse arrangement, but set back from the front axle in order to avoid interfering with the steering angle. The power output was raised to 15 hp at the drawbar and 28 hp on the belt, which was achieved by increasing the cylinder bore to $4\frac{1}{2}$ in and the stroke to $6\frac{1}{2}$.

In all three tractors the maximum engine speed was kept at 1000 rpm.

Transverse engine of the MH 3 tractor.

Front view of the MH 3 tractor.

This was just right, Massey-Harris claimed in their sales literature, as a slower engine speed would mean uneven running and reduced life, while a faster speed would cause excessive vibration and engine wear.

Sales of the MH 1 model appear to have been encouraging, and some were exported, but competition from other manufacturers with more up-to-date models intensified, and the old Parrett design was beginning to lose its appeal. The extra power available in the MH 3 model was insufficient to revive interest and tractor production at the Weston factory ended in 1923.

Production at the Parrett factory had already stopped and that company disappeared from the tractor industry in about 1922.

3 · THIRD TIME LUCKY

The decision to pull out of the tractor market made financial sense for Massey-Harris in 1923. It was a difficult time for the tractor industry everywhere and well over half of the companies which had been making tractors in the boom years of 1916 to 1919 had ceased production by 1925. Some, like Massey-Harris, were able to concentrate on other products, but many more were either taken over or simply ceased to exist.

The problems were caused partly by the economic recession, which drastically reduced the amount of money available for new tractors and farm machinery, and partly by Henry Ford's policy of cutting the price of his Fordson tractors in order to maintain sales volumes. The selling price for the Fordson in the United States in 1918 was $750, but this had been reduced to $395 by 1922.

As the world economy began to expand again in the mid-1920s, sales of Massey-Harris equipment increased rapidly. This was because of the comprehensive product range and an excellent reputation for quality, and the company's strength in overseas markets contributed a major share of the growth with 60 per cent of turnover in the mid-1920s provided by exports and by sales from Massey-Harris factories in France and Germany. The expansion put Massey-Harris among the top six farm equipment companies worldwide.

In spite of this success, the fact that the company had no tractor to sell was once more causing concern. It meant that Massey-Harris was missing the sales opportunities in what had become a major sector of the farm equipment market, and it also meant that the company was not involved in an area of rapidly developing technology.

In 1926 Massey-Harris was once more looking for a way into the tractor market. They wanted to find an American-built tractor which could be imported into Canada for sale as a Massey-Harris product. The advantage of this arrangement was that it saved the development and tooling costs needed to design and manufacture a completely new Massey-Harris tractor, but relying on an outside supplier can also have disadvantages, and some of these had arisen under the previous marketing agreements for the Bull and Parrett tractors.

For their third partner in the tractor industry the Massey-Harris directors chose the J.I. Case Plow Works Co. of Racine, Wisconsin, and negotiations between the two companies began towards the end of 1926.

There were two J.I. Case companies making tractors in the United States at that time, both originally founded by Jerome Increase Case, one of the most successful of the early pioneers of the farm equipment industry. The J.I. Case Threshing Machine Co. was the oldest of the two companies and it was also the biggest. It survived as an independent company until 1965 when a controlling interest was sold to the Kern County Land Co. The Case and International Harvester tractor companies were merged as part of the Tenneco organisation in 1985.

There were no formal links between the Plow Works and the Threshing Machine company, and the two became direct competitors in the tractor market in 1919 when the J.I. Case Plow Works started manufacturing the Wallis tractor. It was the Wallis tractor which interested Massey-Harris in 1926, and this time they had made a good choice because the Wallis 20–30 was well designed and had an excellent reputation.

The directors of the Case company apparently welcomed the approach from Massey-Harris and negotiations should have been straightforward, but instead they became surprisingly complicated. Many of the details were not recorded because discussions were confidential, and there is still some mystery about a situation which at one time threatened to become a major political scandal.

At an early stage in the talks it appears that the Case directors made an offer to buy an interest in the Massey-Harris company. It is unlikely that this idea would have been welcomed by the directors of the Canadian company, but reports about a takeover of Massey-Harris began to appear in the press and these were soon making headlines.

The Massey family still held a controlling interest in the company and nobody else could take control unless a member of the family was willing to sell some shares. Vincent Massey had recently resigned as president of the company in order to free himself to develop a career in politics, and

Wallis Cub Junior with its revolutionary U-frame.

this was a fact which encouraged the public interest. The idea that a senior politician was willing to sell one of Canada's biggest and most prestigious companies to American rivals could have been extremely damaging, and it is hard to believe that anyone as politically and commercially astute as Vincent Massey would seriously have considered such a move.

Instead of the sell-out to Case, arrangements were made for a Canadian business syndicate to buy a controlling interest in Massey-Harris. The company remained firmly under Canadian control and Vincent Massey continued to serve his country in a long and distinguished political career.

Meanwhile negotiations between Massey-Harris and Case continued and, early in 1927, an agreement was signed giving Massey-Harris the right to sell the Wallis tractor throughout Canada and in some areas of the United States. This was not the end of the matter, however, because further negotiations between the two companies ended in 1928 with Massey-Harris buying the J.I. Case Plow Works Co., all rights to the Wallis tractor and the right to use the Case trade name.

The deal cost Massey-Harris a cash payment of $1.3 million plus a guarantee for outstanding bonds worth $1.1 million. Massey-Harris then recovered $700 000 of their outlay by selling their rights to the Case name to the J.I. Case Threshing Machine Co.

This appears to have been a remarkably good arrangement for the Massey-Harris company. It provided the Canadian company with a big

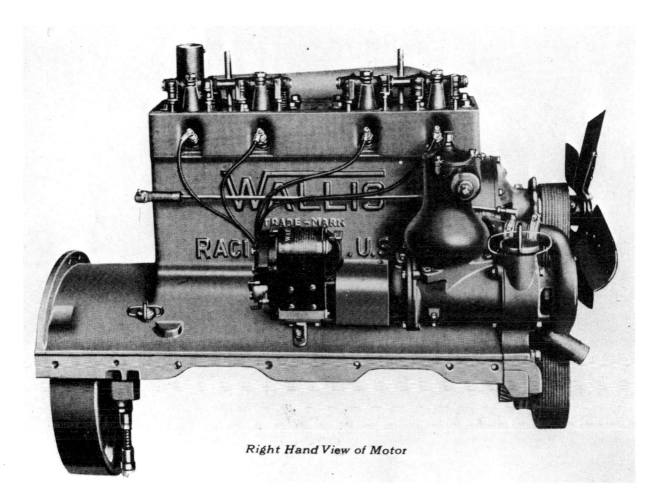

Right Hand View of Motor

The engine used in the Wallis 20–30.

factory in the United States and with a very good tractor which was well established in the immensely important American market, and the cost of all this appears to have been surprisingly small.

As the price paid for the American company was so modest, it casts some doubt on the idea that Case had seriously made a bid for control of Massey-Harris. It would have cost $6 million or more to buy a controlling stake in Massey-Harris, and it is not obvious how such a big investment might have been financed by a company which could be purchased so inexpensively.

One of the distinctive features of the Wallis tractor design, which Massey-Harris now owned, was the curved steel plate enclosing the underside of the engine, transmission and final drive. This was known as the U-frame and had been used on all Wallis tractors since it first appeared on their Cub model in 1913.

The U-frame on the Wallis Cub had covered the underside of the engine and the gearbox, and it was extended under the final drive as well on later models. The idea had been developed by Robert Hendrickson and Clarence Eason in the Wallis company engineering department. Their aim was to provide a more rigid, compact structure which would also protect the tractor mechanism from dirt. This is now recognised as one of the most significant developments in tractor history, but it attracted surprisingly little interest at the time and some manufacturers were still selling tractors with exposed gearing and drive chains, and a steel girder frame for another 10 years or so.

Ford engineers used a further development of this idea for the 1917 Fordson. Instead of the curved steel frame of the Wallis they were able to join the main components of the tractor together in a series of cast-iron structures which provided rigidity and protection from dirt, but was cheaper to manufacture than the Wallis U-frame which was made of high-grade boiler-plate steel.

When the curved steel frame first appeared on the market, the Wallis company had recently been moved into the J.I. Case Plow Works factory. The two continued to share the same premises while operating as separate companies until 1919, when they merged under the Case name but kept the Wallis name for the tractors.

The 20–30 tractor which Massey-Harris took over was powered by a four-cylinder engine designed by Wallis. It had overhead valves and reached its rated output at 1050 rpm. An unusual feature was the gearbox, with only two forward ratios to give travel speeds of 2.75 and 3.3 mph.

In 1927 the Wallis 20–30 tractor was tested at the University of Nebraska where it gave an impressive performance. In the maximum power tests the tractor produced considerably more than the 20–30 rating used by the manufacturer, with a top performance of more than 35 hp in the belt tests and 27 hp at the drawbar. The results showed that almost 77 per cent of the maximum brake horsepower was available at the drawbar, an

Sectioned view of the Wallis 20–30 from a Massey-Harris sales leaflet.

The orchard version of the Wallis
20–30 was sold by Massey-Harris.

exceptionally high figure by 1920s' standards. The excellent drawbar performance was confirmed in 1930 when the tractor appeared as the MH 20–30 at the World Tractor Trials in England. More than twenty wheeled tractors took part in the tests, which included measurement of wheelslip, and only one of the entries achieved better traction efficiency than the 20–30.

Another outstanding feature of the Wallis tractor was its fuel economy, and this was emphasised at Nebraska, where the fuel efficiency figures were among the best achieved in the test series during the 1920s. In the Oxford trials there were only two paraffin-engined tractors which managed to beat the MH 20–30 in the fuel-consumption tests and one of these was an MH 12–20, which gave the Massey-Harris entry an outstandingly consistent record for fuel efficiency against tractors from most of the leading American and European manufacturers.

The MH 12–20 was a smaller tractor announced by Massey-Harris in 1929 as an addition to the range. It retained the Wallis U-frame, and it was styled as a smaller version of the MH 20–30 tractor. The four-cylinder engine was 247.5 in³ capacity compared to the 346 in³ of the 20–30 power unit, and the Nebraska tests showed a maximum belt hp of 24.16 with 18.07 hp at the drawbar. The tractor was equipped with a three-speed gearbox and, for the North American market, it was given a two-plough rating.

Both the MH 12–20 and the MH 20–30 tractors could be equipped with a power-take-off as an optional extra, and there were also road

lights,
were
tomer
firmir
proce
Ma
ments
A r
25–4(
produ
mode
comb
with t
20–3(
Th(
to bu
achiev
which
in mu

A page from a 1930 Massey-Harris sales brochure.

Another feature which was heavily promoted was the Wallis U-frame, and this extract from a 1929 sales leaflet is an example: 'The Wallis Patented U-frame is unquestionably the greatest invention in tractor history. In it are combined light weight, great strength, accessibility and rigidity.' In 1936 it was being promoted as 'The lightest and strongest tractor frame' and this theme was still appearing in a 1940 sales leaflet.

A new look version of the Pacemaker made about 1938.

The next important development in the U-framed tractor series was the arrival in 1936 of the Pacemaker and Challenger tractors. These were both based on the MH 12–20 model, but with the engine developed to produce more power and with a four-speed gearbox. Although both models shared basically the same engine and transmission, the Pacemaker was a standard design and the Challenger was the first Massey-Harris rowcrop model with a high-clearance frame and rear-wheel adjustment from 52 to 80 in.

Engine modifications increased the output to give the new models a power rating of 17 hp at the drawbar and 27 hp on the belt. This was achieved by using a more efficient combustion-chamber design and by changing the carburettor and inlet manifold.

Only 2 years after the Pacemaker and Challenger tractors had arrived on the market, Massey-Harris was announcing new versions of both models. Previously the company had concentrated mainly on engineering changes to improve performance, but this time there was also a completely new look for the Pacemaker with the familiar, angular lines of the old Wallis series replaced by more fashionable rounded styling. Massey-Harris named the new tractor the Streamlined Pacemaker and drew attention to its curves with a bright red paint finish.

The new-look Pacemaker was offered in regular and orchard versions, and the Challenger was retained for rowcrop areas. There was also a

Above:
A new-look Pacemaker exported to Britain.

Right:
Rear view of a Pacemaker showing the offset steering wheel and a power-take-off.

Left:
A Pacemaker ploughing.

Pacemaker illustration used in
Massey-Harris advertising.

choice of engines to burn gasoline only or kerosene, and the gasoline-powered tractors were available with the Twin Power feature which was emphasised in the advertising campaign.

On the Twin Power models there was a small lever on the base of the gear-shift mounting which could be moved into either a high or a low power position. When the low setting was used, the engine speed was limited to 1200 rpm to give a maximum power of 26.3 hp at the drawbar and 36.8 hp on the belt pulley. The high setting was recommended for belt work only and increased the maximum engine speed to 1400 rpm to produce up to 42.1 hp. According to the sales literature, the Twin Power feature gave the new Pacemaker three-plough power at the drawbar but four-plough power on the belt.

Other specification changes included an optional power lift for implements and some minor changes to the engine, including a new magneto, and a more efficient cooling fan and pump.

The list of optional equipment for the new-look Pacemaker included the choice of either a high-ratio or a low-ratio gearbox, but the difference

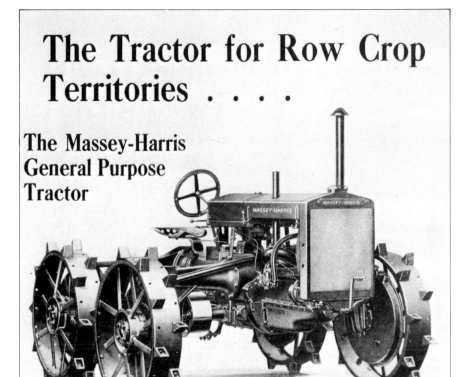

The Tractor for Row Crop Territories . . . .

The Massey-Harris General Purpose Tractor

The Massey-Harris General Purpose Tractor is the result of many years' thought and experiments. It is totally unlike any other row crop tractor and is designed to and actually does replace the horse in farm work. Its ease of starting, hooking to the load, short turn, three speeds, ease of operating, ability to work in narrow quarters so impress the owner that he will not take the time to harness, hitch up, unhitch, and unharness the slow moving horse.

It will successfully operate on any farm under any conditions and with the same load as can be worked with four to six horses. It can be operated from the seat of the drawn implement with the same ease as can be done from the tractor seat. It will pull its load on loose or wet soil and every ounce of its weight is balanced over its four traction wheels when a normal load is pulled. This tractor drives from all four wheels —no load to push—no loss of drawbar pull—no waste of fuel, increases the life of the tractor, its usefulness and lessens the cost of maintenance.

**FOR MORE COMPLETE INFORMATION
SEE THE MASSEY-HARRIS DEALER**

between the two was surprisingly small. In the low-ratio version the maximum forward speeds ranged from 2.1 mph in bottom gear to 7.5 mph in fourth gear, and these were increased to 2.4 and 8.5 mph for customers who chose the high-ratio gearbox.

Within 10 years of buying the Case Plow Company, Massey-Harris

Action pictures of the General Purpose tractor from the Massey-Ferguson archives.

was established among the top five companies in the tractor industry, and this success was almost entirely due to the results achieved with tractors based on the former Wallis design.

An attempt to sell a completely different type of tractor, which had not been developed from the Wallis, was much less successful. This was the General Purpose model announced by Massey-Harris in 1930. The special feature of this model was the use of four-wheel drive through equal-size wheels, an arrangement which is now recognised as the most efficient way to achieve maximum traction with a wheeled tractor.

In 1930 farmers who wanted extra pulling power chose a crawler tractor or an ordinary two-wheel drive model with a bigger engine. Few manufacturers had attempted to cope with the extra mechanical complexity of four-wheel drive, and the equal-size wheel version was almost unknown. The General Purpose model of 1930 was an imaginative attempt to design a more efficient tractor and it failed, basically because it arrived on the market about 25 years too early. Few farmers appreciated the advantages of the four big driving wheels so the demand was small; also the technical know-how for combining drive and steering systems was fairly primitive and the tractor was clumsy and hard to manoeuvre.

Massey-Harris chose a Hercules engine for the General Purpose tractor. This was a surprising arrangement, as the Hercules engine was 226 in^3 capacity and Massey-Harris might also have used their own 247.5 in^3 engine already in production for the MH 12–20 tractor.

The Hercules engine provided a maximum of 24.8 hp on the belt and 19.9 hp at the drawbar operating at 1200 rpm during its Nebraska tests. The transmission gave three forward speeds of 2.2, 3.2 and 4 mph, with a three-plate clutch. The long list of optional extras included a set of road lights, an electric start kit, a set of extension controls so the tractor could be operated from the seat of a trailed implement, plus a power-take-off. An orchard model with fenders to protect the top half of each wheel was also offered and a power-lift kit was available later.

Massey-Harris designed the General Purpose as a rowcrop tractor with 30 in of clearance under the axles and a range of wheel settings from 48 to 76 in. Horses were still widely used for inter-row work and they still attracted such intense loyalty that the idea of replacing them with a tractor had to be expressed with the sort of diplomacy which appeared in the General Purpose tractor sales leaflet.

'It is totally unlike any other row crop tractor – it was designed to ACTUALLY REPLACE HORSES in the corn and cotton belt under ANY soil conditions'. But, the leaflet tactfully added, 'We love horses and do not wish you to think otherwise. However, we know the farmer must lighten his operating cost and when he does, his answer will be – the Massey-Harris General Purpose Tractor.'

Massey-Harris also had to explain the benefits of the unorthodox design to potential customers. The 1930 sales leaflet explained that it was the only tractor 'so constructed as to flexibility that each of the four drive wheels provides constant contact with the ground and assures balanced traction at all times and under all conditions'.

Rowcrop work does not usually provide the most demanding conditions for tractors and is not the obvious way to make use of four-wheel-drive pulling power, but farms with steep land can benefit from the stability and wheelgrip of four driven wheels, and this was another market sector Massey-Harris tried to develop for the General Purpose tractor.

In 1936 a new version of the General Purpose was announced in a bid to revive the disappointing sales figures. The engine was now available in both gasoline and kerosene versions. Massey-Harris had developed a new type of vapouriser for the engine, with an adjustment which could be reached from the tractor seat to vary the heating effect from the exhaust in order to achieve efficient combustion. The new model, called the Four-Wheel Drive, also had extra wheel settings, and rubber tyres and an industrial version were available as options.

These improvements did little to improve the sales figures and Massey-Harris abandoned their first venture into four-wheel drive and the first tractor which had been designed within the company.

4 · A NEW GENERATION

Apart from the four-wheel drive General Purpose model, all the tractors sold by Massey-Harris up to 1938 were based on designs bought in from other companies.

This situation changed with the arrival of a new generation of Massey-Harris tractors, starting with the MH 101 model which was announced in 1938 and remained in production until 1947 at the old J.I. Case factory at Racine. The 101 was offered in standard, Super, Senior and Junior versions, and with Twin Power as an option. There were also the confusingly numbered MH 102, 201, 202 and 203 models, plus the smaller MH 81 and 82 to provide the Massey-Harris line up through the war, with the addition of the little General tractor which was bought in and made only a brief appearance with Massey-Harris.

Although this was the first full range of Massey-Harris designed models, none was powered by a Massey-Harris engine. The MH 101, 101

Harvesting wheat in Ontaric with an MH 102 tractor.

Super and the 201 were equipped with Chrysler six-cylinder gasoline
engines and the other models all used four- or six-cylinder Continental
power units.

It was the Chrysler engines which attracted most attention at the time.
These were designed mainly for trucks, and it was still considered unusual
to specify an automotive-type engine in a tractor, and something of a
luxury to have a six-cylinder power in a tractor with a two-plough rating
like the MH 101. Most of the other big tractor companies in North
America preferred to make their own engines, but the policy of relying

on outside suppliers was one which Massey-Harris turned to advantage in their sales promotion.

The Chrysler engine offered all the refinement associated with automotive engineering, plus the smoothness and quietness of six cylinders, Massey-Harris claimed. As the engine type was popular for trucking operations, the Massey-Harris sales leaflet could claim it was backed by '12,000 million miles of performance'. Another benefit stressed in the leaflet was that the combination of Chrysler and Massey-Harris dealers provided the most extensive service network of any tractor make.

An electric starter motor was included in the MH 101 standard specification, and in 1938 this was still sufficiently unusual for Massey-Harris to describe it as 'the first tractor made by a full line farm equipment company with a self-starter as standard equipment'. For those who doubted the value of such a luxury, the Massey-Harris publicity department explained that the operator would no longer be tempted to leave the engine idling to avoid the chore of cranking it back to life again. This could reduce gasoline consumption by up to 10 per cent, the company claimed, but without explaining where they had obtained their evidence for such a big saving.

The Twin Power feature had been inherited from the Pacemaker. The

MH 81 tractors built for the Royal Canadian Air Force.

The General tractor in the Ontario
Agricultural Museum, Canada.

two-plough MH 101 with Twin Power was claimed to give three-plough
tractor performance on the pulley. A variable speed governor allowed a
maximum 1500 rpm from the engine while the first, second or third gears
were engaged, but was automatically reset to allow 1800 rpm and a 20 per
cent increase in power for operating the pulley or when the top gear was
engaged.

The MH 101 and its Junior and Senior derivatives were the mainstay of
Massey-Harris tractor business through the war years, while the rest of the
tractor range made only a small contribution to the sales totals. The other
models included the MH 81 series, available from 1941, which was
bought by the Royal Canadian Air Force for towing work on airfields.

Another of the wartime models was the The General, described by
Massey-Harris as 'the new little tractor that does a big job of supplying
dependable, economical power for small farms'. It was brought in to give
the Massey-Harris company and its dealers a model to sell in the expanding market for small tractors. All the major companies in the industry
were moving into this sector and Massey-Harris, with no previous experience of small tractor design, made an agreement to buy in the tractor for
distribution through Massey-Harris outlets in Canada and parts of the
United States.

The General was built by the Cleveland Tractor Company, an old
established American manufacturer specialising in the tracklaying models
with which they achieved considerable success under the Cletrac trade

tion and bought in The General. This situation changed in 1946 when the first of a full range of new models arrived on the market. Some of these were powered by engines supplied by Continental, but two of the most important models, the MH 44 and 55 tractors, were both equipped with engines designed and made by Massey-Harris.

By the end of the 1940s, when the new line was complete, there were Massey-Harris models ranging from the little Pony tractor, with a rated output of 10.3 hp on the belt, up to the MH 55 which was given a 52 hp rating for the gasoline version. In between were the 20 and 30 series, plus the highly successful MH 44. The MH 44 was undoubtedly the company's star performer at that time, with production from the Racine factory peaking at almost 20 000 in 1951, plus a further 4600 of the British-built MH 744 version built initially in Manchester and later in Kilmarnock, Scotland. Production of the MH 44 and MH 744 in the 3 years from 1950 to 1952 was more than 50 per cent of the company's total tractor output.

The Pony was also a success, but only in France where production from the Marquette factory, with the model number prefixed by 8, reached a

The spring-mounted seat of the MH 44 and MH 744 tractors.

Above right:
Half-track conversions gave the MH 744D extra traction.

Above left:
The MH 744 was the British version of the MH 44.

Left:
The British version of the MH 44 was equipped with a Perkins engine.

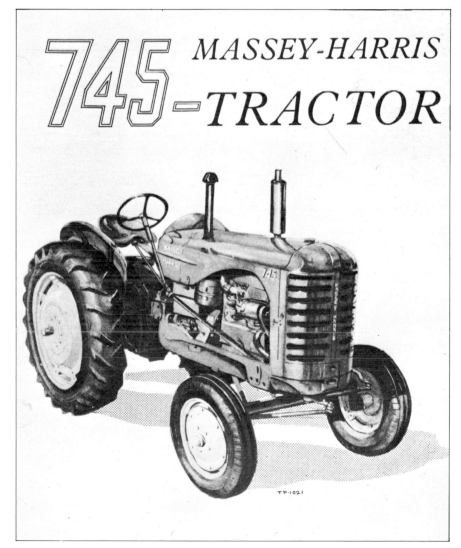

745 MASSEY-HARRIS -TRACTOR

TP-1021

Publicity material for the MH 745.

A French-built Pony tractor with a mid-mounted mower.

Left:
Massey-Harris publicity illustration of a Pony in action.

Pony 820 tractor pictured at the Paris show.

maximum of more than 14 000 in 1958. The version which made the biggest impact was the 820, manufactured from 1957 and powered by a diesel engine supplied by Hanomag of Germany. More than 100 000 Pony tractors were built in France before production ended in 1965.

Pony tractors were also built at the Woodstock factory in Canada, but the production figures there remained well below those achieved at Marquette, and this is one indication of the weakness of the Massey-Harris range in the lower end of the power scale. One reason for the poor sales performance was the fact that Massey-Harris had previously concentrated on medium-power models while most of the other big names in the American and Canadian markets had been selling one-plough tractors for years.

Above and opposite:
Pony tractors pictured in Canada.

54

Another important factor was the superiority of the small Ferguson System tractors from Ford and, later, from the Ferguson factory in Detroit. The management at Massey-Harris was increasingly aware of the problem and they tried to bridge the technology gap with the Depth-O-Matic hydraulic-lift system. This first appeared as an option on the MH 20 tractor in 1949 and was later extended to other models including the mid-range MH 30 in 1950. The Massey-Harris design lacked draft control and the hydraulics were an external addition, which meant that the Depth-O-Matic was a poor substitute for the full Ferguson System.

Another attempt to match the Ferguson System was the experimental Pitt tractor of 1953. Professor Arnold Pitt, Associate Professor of Physics at the University of Toronto, was put in charge of a new research department at Massey-Harris which had been formed to augment the Engineering Division. The department was given the job of developing a new implement attachment-and-control arrangement which would be as effective as the Ferguson System. For security reasons, the experimental prototype was referred to as the Pitt tractor.

The Pitt tractor had been dismissed as a failure in July 1953 when the Massey-Harris president, James Duncan, was making an assessment of the advantages and disadvantages of the proposed association with the Ferguson organisation. This included the admission that Massey-Harris had nothing to match the Ferguson System, and that the MH 22 and the Mustang – a 24 hp 20 series model – with Depth-O-Matic could not compete with the new Ford tractor equipped with a Ferguson-type implement attachment system. To turn down the opportunity of bringing the Ferguson System into the Massey-Harris range 'would be tantamount to abandoning the 2-plow tractor field', he concluded.

5 · THE FERGUSON SYSTEM

Harry Ferguson's system for attaching implements to tractors was one of the most important developments in the history of power farming. It made a major contribution to efficiency and to safety, and it is still the most widely-used arrangement for implement attachment and control more than 50 years after the first Ferguson System tractors arrived on the market.

The man who spent much of his life developing and marketing the system was born in what is now Northern Ireland on 4 November 1884. His home was a 100-acre farm at Growell, County Down, where Harry was one of eleven children.

As he grew older he was expected to do an increasing share of the jobs

Harry Ferguson.

on the farm, and this experience of the drudgery of manual work and the slow pace of farming with horses encouraged his decision to leave home at the first opportunity.

The opportunity came in 1902 when Harry was still 17 years old. His elder brother, Joe, invited Harry to join him in a small workshop he had started in Belfast doing repair work on bicycles and cars. Harry had already developed a keen interest in anything mechanical and he accepted the offer and started his new job with enthusiasm. He was quick to learn and showed a particular aptitude for servicing and repairing temperamental car engines with a skill and thoroughness which helped to establish and expand the business.

One of Harry Ferguson's early interests was in motor sports and he became an enthusiastic competitor in motor-cycle racing and trials events and, later, in sports-car racing. He achieved considerable success, helped by his determination and, no doubt, by his skill and thoroughness in preparing the motor-cycles and cars for each event. He was quick to appreciate the advertising value of winning races and his successes helped to publicise the Ferguson brothers' business which was now a well-established garage.

Motor racing remained one of Harry Ferguson's lifelong interests. During the 1920s he was prominent in the campaign to establish a major motor-sport event in Northern Ireland, and the result was the famous series of Ulster Tourist Trophy races which began in 1928. One of his last major projects, which was started only a few months before his death in 1960, was to plan a new Formula 1 racing car which later achieved some notable successes with Stirling Moss at the wheel.

Another of his interests was flying. This started in 1908, when aviation was still very much a novelty, and provides further evidence of Harry

The aircraft in which Harry Ferguson made his first flight in 1909.

58

Aircraft at the Ferguson garage in Belfast, Ireland.

Ferguson's mechanical skills, determination and personal courage. He designed and built his own aircraft and made his first successful flight in December 1909. This was the first time an aircraft had flown in Ireland, and it also made Ferguson the first person in Britain to design and fly his own aircraft.

Harry Ferguson built two or three more aircraft and made many more flights, crashing several times in the process, before other activities claimed his interest.

In 1911 Harry Ferguson decided to leave his brother's business and start his own garage in Belfast. Establishing the new business was one of the new interests which helped to end his career as an aviator. Another development was his marriage in 1913, and this may have been another reason for giving up such a hazardous pastime as flying home-made aircraft.

Up to this time Harry Ferguson's only close association with agriculture had been helping on the family farm, but World War 1 brought him back into contact again and started his interest in farm mechanisation.

During the war, the Government organised a major campaign to in-

crease food production from British farms as German U-boats almost
succeeded in preventing supplies reaching Britain. The most effective way
to increase production quickly was to make use of tractor power to
plough up large areas of pasture to grow cereals and potatoes. The food
campaign included Ireland, still under British control, and tractors were
used in substantial numbers for the first time.

Harry Ferguson recognised the situation as a business opportunity and
took the agency for the Waterloo Boy Model N tractor. This was built
by the Waterloo Gasoline Engine Co. of Waterloo, Iowa, and was imported into Britain, where it was known as the Overtime and became one
of the most popular tractors during the war. The Waterloo company was
later taken over when John Deere decided to move into the tractor
market, and the Model N was the first tractor manufactured by John
Deere.

The Overtime gave Harry Ferguson his first experience of power farming as he helped to supervise ploughing demonstrations as part of the sales
campaign. Although there are no records to show how many Overtimes
he sold, the demonstrations must have been effective because they attracted the attention of the Irish Board of Agriculture.

In 1917 Harry Ferguson was asked to tour Ireland on behalf of the
Board to help improve the standard of tractor operation. At that time,
while horses were still the main source of power, few farmers had any
experience of using tractors and the results were often poor. Ferguson's
job was to arrange local demonstrations and to visit owners in order to
show how to make the best use of tractor power.

Harry Ferguson made the tour during March, travelling with his assis-
tant, William Sands, a member of the staff at his garage. The tour pro-

vided Ferguson with an excellent opportunity to see different tractor makes in action and to study the way the engine power was utilised for jobs such as ploughing. What he and Sands saw helped to convince them that simply relying on the weight of the tractor to provide sufficient traction to drag the plough through the soil was inefficient, and this was probably the first stage in the development of what eventually became the Ferguson System.

A patent application covering an improved plough design was filed by Ferguson on 12 September 1917, the first of a long series of patents (see Appendix 1) which trace the various stages in the development and improvement of the Ferguson System.

During November 1917, Harry Ferguson gave the first of a series of demonstrations to show his new plough in action. It was called the Belfast plough and was made in his garage, and it included some basic ideas which were to remain an essential part of Harry Ferguson's approach to tractor and machinery design. One important feature was the way the plough was hitched to a point underneath the tractor and ahead of the rear axle, so that hard soil conditions produced a downward pull on the tractor to aid traction.

The plough design had been simplified to reduce the number of parts and to minimise the weight, and it was designed for the Eros tractor, a lightweight adapted from a Ford Model T car by means of a conversion kit made by the Staude company of St Paul, Minnesota. The Eros kit was one of several makes designed to give some of the millions of ageing Model Ts a new lease of life on the land as a low-cost alternative to a tractor. The cheapness and the light weight of the Model T conversion were features which would have appealed to Harry Ferguson much more than the heavy, outdated design of the Overtime tractor.

A Ferguson plough on a Model T Ford car with an Eros conversion.

Although the Eros attracted considerable interest, this came to an abrupt end with the arrival of Henry Ford's new Model F Fordson tractor from Dearborn, Michigan. This tractor was backed by the massive resources of the Ford Motor Company and was manufactured in large numbers to sell at prices which were so low that many other manufacturers were forced out of business. Harry Ferguson had to abandon the Belfast plough and produced a new design to suit the Fordson tractor.

A new design feature in the plough for the Fordson was the Duplex hitch. This consisted of two struts, which helped to keep the plough in the ground and also reduced the risk of the tractor overturning if the plough hit an obstruction. Accidents with early Fordson tractors had killed 136 people in the United States by 1922, according to one estimate. The accidents occurred when a plough hit an obstruction thus putting a sudden shock load through the drawbar and causing the lightweight tractor to rear over backwards, giving the driver little chance to escape.

Harry Ferguson's new plough was manufactured in the United States by the Sherman brothers of Evansville, New York, while Harry Ferguson remained in Northern Ireland to continue the development work on an improved hitch system. He and William Sands designed a hydraulic control arrangement and converted the Duplex hitch into a three-point linkage.

As the design progressed, it became increasingly evident that a new tractor would be needed, together with a set of implements in order to market the system, and Ferguson started to look for a manufacturer who

Harry Ferguson (*left*) watches a 1921 demonstration with one of his ploughs on a Fordson tractor in Lincolnshire, England.

would put the equipment into production. Several companies showed some interest in the equipment, but the market for tractors and machinery in the late 1920s was still depressed and there was little enthusiasm for a project involving a big investment.

In order to prove the superiority of his implement attachment-and-control system Harry Ferguson decided to build a special tractor which could be used to demonstrate the system. The result was his Black Tractor, completed in 1933 and now in the Science Museum in London.

The Black Tractor, with a glossy black paint finish, was the first Ferguson System tractor and it achieved its objective of attracting a manufacturer. The person with sufficient faith in Ferguson and his ideas, and with the resources to put a Ferguson tractor into production, was David Brown, later Sir David, who had recently become managing director of the family gear-cutting business at Huddersfield, Yorkshire. His company had been given the contract to cut a special set of gears for the Black Tractor, and this initial contact was followed eventually by the production agreement Ferguson wanted.

Production started in 1936. David Brown had formed a company under his own name to build the tractors in premises rented from the gear company, and Ferguson and his backers formed a marketing company responsible for selling the tractors and implements. The tractors carried the Ferguson name above the radiator, but a small plate just above the starting handle bore the name of David Brown Tractors.

The tractor model was known as the Ferguson Type A, but is often called the Ferguson-Brown. The design was based closely on the Black Tractor, but the paint colour was changed to battleship grey and the Hercules engine which powered the prototype tractor was replaced by an 18–20 hp Coventry Climax E for the first 500 production tractors, and by a 2010 cc David Brown unit for the rest of the production run.

Harry Ferguson Ltd launched the tractor at a public demonstration in May 1936 and a press report at the time claimed 'an extraordinary de-

The Ferguson name at the top of the radiator and — . . . .

. . . . — the David Brown name at the bottom.

1 MH 1 tractor built at the Weston
Factory, Toronto, in 1918.

Right:
3 Sawyer-Massey 20–40 tractor.

2 The Sawyer-Massey name on a
1920 steam traction engine.

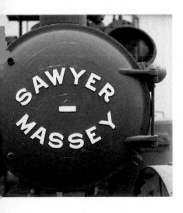

4 A Wallis 20–30 sold by Massey-Harris in about 1929.

5 MH 25 in the Ontario Agricultural Museum, Canada.

6 A late version of the Massey-Harris General Purpose tractor.

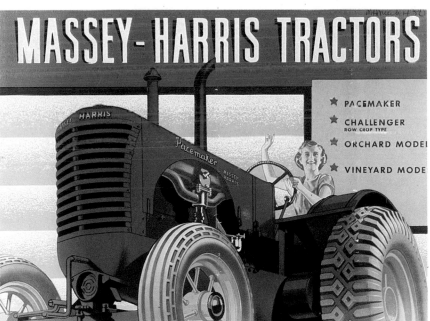

7 The cover of a Pacemaker tractor brochure.

9 Twin Power version of the Pacemaker.

Left:
8 A Pacemaker built in about 1938.

10 MH 101.

11 The cover of an MH 101 sales brochure.

12 The General was sold by Massey-Harris in 1940–41.

13 MH 102 built in 1942.

14 A Canadian-built Pony tractor.

Opposite above:
15 MH 30.

Opposite below:
16 The 55 tractor was at the top of the new Massey-Harris range in 1946.

Right:
17 The MH 44 was known as the MH 744 in Britain.

Below left:
18 Ferguson Type A front-end detail.

Below right.
19 Ferguson Type A tractor No. 1.

20 Ford 9N.

Left:
21 The first of the Coventry-built TE tractors.

Opposite above:
22 Ferguson TE with original loader.

Opposite below:
23 Prototype Standard tractor.

Left above:
24 TO 35 diesel tractor from Detroit, United States.

Left below:
25 Ferguson 40.

Right:
26 The grey and gold Ferguson 35 tractor from Coventry.

Opposite above:
27 MH 50.

Opposite below:
28 The MF 95 Super, one of the big 90 series tractors.

Right:
29 MF 1200 with articulated steering.

Below:
30/31 Attractive advertising by Landini.

32 A Landini tractor working in
Italy.

33 French-built 2000 series tractor
available with electronic linkage
control in 1985.

Tractor assembly line at Huddersfield, England.

Coventry Climax engine fitted on the early Type A Fergusons.

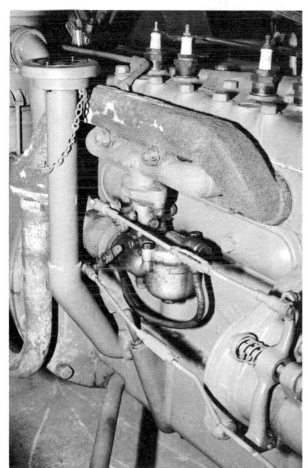

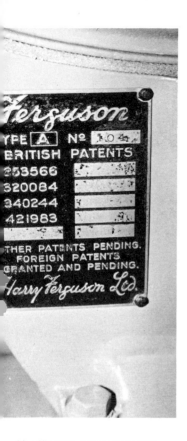

Identification plate on Ferguson
Type A No.104.

Right:
**Ferguson plough built for the Type
A tractor.**

Above left:
**Mudguard design on the early Type
A Ferguson.**

Left:
**Ferguson Type A No.104 equipped
with road bands.**

mand' for the new tractor. This was followed by a highly optimistic claim
in the June issue of *Implement & Machinery Review* that 'over a thousand
have already been sold'.

There was certainly plenty of interest in the tractor, which could easily
outperform bigger, more powerful competitors at public demonstrations,
but the sales figures were much lower than the press reports had suggested. One problem was price, with the Fordson on offer at £140,
against £224 for the Ferguson, plus a further £28 for each implement.
Harry Ferguson tried to overcome this barrier by listing the tractor at
£178 plus £46 for the hydraulics, but this does not appear to have helped
the situation significantly and unsold tractors were soon accumulating at
the Huddersfield factory.

Improvements in 1937 included optional rubber tyres and an orchard model with a tighter turning circle, but the sales graphs refused to turn upwards and the relationship between the production and the marketing companies started to deteriorate. David Brown wanted more design improvements, including extra power and four gears, while Ferguson complained of quality control problems in the factory.

In 1938 the friction between the two sides increased. David Brown decided to go ahead with plans for the design changes he wanted, and Harry Ferguson set sail for the United States to meet Henry Ford.

The meeting took place on land near Henry Ford's Fair Lane mansion on the outskirts of Dearborn, and had been arranged to provide Ferguson with an opportunity to demonstrate his system of implement attachment and control. Henry Ford, like Harry Ferguson, had been brought up on a small family farm, but had left home to develop his mechanical skills and interests. Although cars had contributed most to the extraordinary growth of his company, Ford remained deeply interested in tractor development and quickly appreciated the improved performance of the Ferguson System compared with a conventional tractor and plough.

The demonstration ended with the famous handshake agreement. Henry Ford agreed to provide the resources to put a new Ferguson System tractor into volume production, and Harry Ferguson undertook to form a new company to market the tractors together with implements

Steering modification on the Ferguson Type A orchard version.

Above right:
Power-take-off drive from the belt pulley on the Ontario Agricultural Museum 'Ferguson-Brown'.

Right:
Harry Ferguson and Henry Ford at the meeting where they made their famous 'handshake' agreement.

to be sourced from outside suppliers. It was also agreed, apparently, that either partner could terminate the agreement at any time.

It was an arrangement which involved the patents representing 20 years of work by Ferguson and his team, and it also involved a multi-million-dollar investment for the Ford organisation. Curiously the agreement was never witnessed or recorded in any way, which was evidence of the mutual trust and confidence between Ferguson and Ford, but eventually caused immense problems when legal experts tried to decide exactly what had been agreed.

Work on the new tractor project started immediately. Ford engineers dismantled the Ferguson tractor to familiarise themselves with the hydraulic system, while Ferguson returned briefly to England to deal with various business matters before returning to Dearborn with his family and his engineering team.

The Ferguson-Ford deal left David Brown with a stock of unsold Ferguson A tractors and with the freedom to make the design changes Ferguson had always refused. An announcement in the February edition of *Implement & Machinery Review* stated that: 'Owing to the maturing of plans to cater for the export market, it has now been considered necessary, we learn, to segregate the Ferguson and Brown interests so that no confusion should arise in certain markets'.

Also in this announcement was a suggestion that there would be a licencing arrangement so that the David Brown company could continue to manufacture and sell the Ferguson tractor. Instead the David Brown team concentrated on developing a completely new model, the VAK 1, with several improved features including a power-take-off which was designed as part of the tractor instead of the external attachment used on the Ferguson. The new David Brown model, without Ferguson System hydraulics, was announced in July 1939, while the company was still trying to sell off the old Ferguson A tractors in September at the reduced price of £198.

The new Ford tractor was the 9N, correctly known as the 'Ford tractor with Ferguson System'. It was demonstrated before an audience of 500 guests at the Ford farm at Dearborn on 29 June 1939 when Harry Ferguson took a turn at the wheel. The tractor, bearing the oval Ford badge over a small Ferguson System plate, was painted Ferguson's favourite shade of grey.

Demonstrations organised by Harry Ferguson were always prepared and carried out with great care to achieve the maximum effect, and the 9N demonstration at Dearborn included evidence of the Ferguson method. A tractor and plough were set to work in a fenced enclosure measuring only 20 ft by 27 ft, showing that the driver could plough all the enclosed area and leave no wheelmarks. Another demonstration technique, used later when Ferguson brought a 9N to Britain, was to plough at 6 in depth with two 10 in furrows, while two of the four spark-plug leads were disconnected.

Soon after the new Ford tractor was launched there were restrictions on

Harry Ferguson (right) at a demonstration of the Ford 9N tractor on an experimental farm in Ireland.

production because of the wartime shortage of raw materials, and the production line was stopped completely for a short time during 1942. Because of the production problems some of the tractors built during World War 2 were produced with steel wheels instead of the standard-specification rubber tyres, and with magneto ignition to overcome the shortage of starter motors and batteries, and these tractors were known as the 2N model.

In spite of the production problems the tractor was popular and, when the wartime restrictions were lifted, the sales figures increased sharply to more than 70 000 in 1946, the last full production year. The success of the 9N and its rapidly increasing share of the small tractor market in the United States and Canada caused problems for other manufacturers and some, including Massey-Harris, tried to develop their own systems of implement attachment but with little success.

As soon as the 9N tractor was in production and his sales organisation was operating satisfactorily in the United States, Harry Ferguson returned home with plans to arrange for a Ferguson System tractor to be built in Britain. He had assumed that the Ford factory in England would stop building the Fordson, a direct and ageing descendant of the 1917 Model F design, and put the 9N into production instead. Henry Ford appears to have supported this idea but the directors of the British company thought otherwise. It would have been difficult to launch a new model while the factory was geared up for wartime production of the Fordson, and they may also have been encouraged by the way the Fordson had retained its popularity previously when Harry Ferguson had been selling his Type A tractor.

After the war, when the Ford factory at Dagenham might have changed over to building a Ferguson System tractor, the company announced the E27N, another derivative of the old Fordson design. Harry Ferguson was forced to make alternative arrangements, and his new British partner was Sir John Black, chairman of the Standard Motor Company which had a large factory at Banner Lane, Coventry.

The factory had been used for assembling aircraft engines during the war and there was plenty of space available for tractor production. The agreement with the Standard company followed the pattern which had already been established with David Brown and the Ford company, with Ferguson forming an organisation to market the tractor, and also retaining control over engineering development, while his partner was responsible for production.

In 1946 the first of the new Ferguson TE series tractors came off the Banner Lane production line. The styling of the new model closely resembled the Ford 9N, but with a four-speed gearbox. The power unit for the first 2 years of production was supplied by Continental, with over-

The first of the Ferguson TE series.

head valves and a maximum output of 24 hp from 1966 cc capacity. This was replaced in 1948 by an engine of similar power made by the Standard company.

The TE became one of the most popular tractors ever built, and Harry Ferguson was probably delighted when his new tractor pushed the Fordson off the top of the British sales charts.

Meanwhile, in the United States, the partnership between Harry Ferguson and the Ford Motor Co. was coming to an end. Ferguson's interests had remained reasonably secure while Henry Ford was in control of his company as both men shared many objectives and ideas, and there was a great deal of mutual respect between them. Henry Ford regarded the

The radiator of the first British TE tractor built at Coventry.

This prototype design for a bigger 'Ford-Ferguson' tractor never went into production.

74

arrangement with Ferguson as a continuing situation and he authorised development work on a new Ford-Ferguson tractor. The new model would have been more powerful than the 9N, and the probable plan was to manufacture both models after the war, covering the small and medium sectors of the market with what could have been a formidable tractor range.

The new model was developed under the terms of the original Ford-Ferguson agreement, carrying both the Ford and Ferguson names. At least two prototypes were built in 1944 while Ferguson was in the United States during the war, but that was the end of the project. In 1945 Harry Ferguson returned to Britain and Henry Ford handed over the control of his company to his grandson, Henry Ford II.

When the new president took over, the company was in a difficult financial situation and there were serious problems to be resolved. The new tractor model was probably shelved at this stage, and Henry Ford II also took a close look at the manufacturing and distribution arrangements for the 9N and decided that the situation was not ideal. The 9N had been losing money and the new president decided to end the marketing arrangement with Ferguson so that his company would have total control over sales as well as production.

Several attempts were made by the Ford company to negotiate a new agreement which would give Harry Ferguson a minority interest in a new distribution company. When Ferguson refused, the management at Ford decided to act without his agreement. In November 1946 plans for an improved version of the 9N tractor were announced and a new sales company, called Dearborn Motors, was set up to market the new tractor. The new arrangement would become effective in June 1947, and this

Ford 9N tractor pictured in Lincoln-shire, England.

allowed Ferguson a further 6 months to operate under the previous agreement with the 9N.

This put Harry Ferguson in an extremely difficult situation, with a large distribution company which would soon have nothing to sell, and with the prospect of losing a substantial amount of royalty income when the 9N was phased out. To add, in Ferguson's view, insult to injury, the 8N tractor launched by Ford in 1947 to replace the 9N was equipped with full Ferguson System hydraulics, including patented features used without authorisation.

Harry Ferguson reacted with typical determination and courage. He started a legal action against the Ford Motor Company for sums eventually totalling $340 million, alleging patent infringements and loss of business due to the changed marketing arrangements. He also started work on the reconstruction of his American business, establishing a new factory in Detroit to manufacture Ferguson tractors for his distribution company to sell. The American-built Ferguson was the TO 20, based on the British TE 20 design, and 25 000 TE 20s were shipped across from the Coventry factory to keep the American Ferguson organisation in business until the TO version was available from Detroit.

Harry Ferguson at the controls at Claridges Hotel, London, in 1948.

The 500 000th Ferguson is in the Massey-Ferguson Museum in Warwickshire, England.

Winston Churchill, seated, watches a Ferguson tractor working on his estate at Chartwell, Kent, England.

The legal action dragged on through more than 4 years to become one of the most complex and costly lawsuits of its kind. It would have been possible to bring the case to a close much more quickly if Ferguson had shown greater flexibility and willingness to accept a compromise, but he pursued the matter with tenacity, fighting for what he regarded as a matter of principle involving the patents which were his life's work.

Eventually, in April 1952, a settlement was agreed and the legal action came to an end. There was no compensation for loss of business allegedly caused when Dearborn Motors took over the distribution. This was probably because of the immense success Harry Ferguson had achieved in establishing his own supply of tractors to distribute, making it difficult to sustain a claim that the Ford action had caused lasting harm to his business. The claim for patent infringement with the 8N tractor design was proved, and Ferguson accepted a payment of $9.25 million in settlement plus an agreement that the design would be modified to avoid further infringement.

In financial terms the settlement was a poor return for the time and money the action had cost Ferguson and his company, as legal costs and

A creeper gearbox added to a 1947 TE 20.

TE 20 tractor in the Antarctic.

payments to other companies involved in the case took almost half the sum agreed, but in other respects the outcome was the victory which Ferguson claimed it to be. The Ferguson System patents were upheld under a most thorough and searching examination, and the lawsuit attracted immense publicity which was generally favourable to Ferguson and was of positive value to his business interests.

With the lawsuit behind him, Ferguson was able to concentrate on his business interests and on the Price Reducing System which was the economic plan he promoted at almost every opportunity.

The Ferguson economic theory was based on his conviction that inflation was at the heart of many of the world's most serious evils, causing unemployment and the widespread poverty which encouraged Communism. The answer to inflation was to reduce prices, he claimed, starting with agriculture and using more efficient mechanisation to bring down the cost of food.

The Price Reducing System became a crusade which Ferguson followed with immense determination. He tried to persuade leading politicians in the United States and Britain to adopt his ideas, he paid for

Steerable skis developed for the Antarctic.

newspaper advertisements to publicise the System and he encouraged other industrialists to put the plan into action. His campaign had little effect, excepting in his own companies where the Price Reducing System undoubtedly held down profit levels and reduced Harry Ferguson's personal income.

Another factor which helped to reduce the profit levels of the Ferguson companies was the sharp fall in tractor sales which followed the end of the Korean war. The increased demand for tractor power soon after the war started had helped to establish the Ferguson TO in the North American market, and the Ferguson factory in Detroit was seriously affected as the wartime sales boom passed.

Harry Ferguson's response to the problems facing his American company was surprising. In 1952 he decided to sell the Detroit-based company and the buyer he had in mind was Massey-Harris. Soon after talks between Harry Ferguson and the Massey-Harris president, James Duncan, started, their scope was widened because Harry Ferguson decided to sell the whole of his tractor business.

Exactly why he reached this decision is not clear. He may have been dissatisfied by the progress made in establishing the Ferguson System on a

The steering wheel was replaced by two hand-operated levers for the tracked versions of the Antarctic tractors.

worldwide basis and believed an association with Massey-Harris would help to achieve this. Another possibility is that he wanted to make additional time and money available to develop his recently-acquired interests in the motor industry. He may also have been aware, at a time when his health was causing concern, of the need to ensure that the ideas and developments he had pioneered would be in safe hands for the future.

Colin Fraser's account in *Harry Ferguson – Inventor and Pioneer* suggests that Ferguson became so keen for Massey-Harris to take over his tractor companies that the deal was agreed with a minium of negotiation and for less than the sum Massey-Harris might have been prepared to pay. At one stage a difference of $1 million was settled by the toss of a coin – it was Harry Ferguson's idea, and he lost.

A public announcement in August 1953 presented the arrangement as a merger but the reality was a takeover. Ferguson received Massey-Harris shares worth about $16 million, the honorary position of chairman, and control over tractor design. It was also agreed that the name of the company would be changed to Massey-Harris-Ferguson, but this was later simplified to Massey-Ferguson.

Absorbing the Ferguson companies into the Massey-Harris organisa-

tion caused considerable discord, especially as many of the Ferguson staff had been unhappy about the deal. There was also discord soon after Harry Ferguson had established himself in his new position as chairman and engineering adviser.

One of many problems was the fate of the LTX tractor, the prototype for a bigger version of the TE and TO series which Ferguson engineers had completed just before the 'merger'. This was to be powered by a 190 in³ petrol-paraffin engine, with a 224 in³ diesel offered as an alternative. Both engines were designed by the Ferguson team and were based on the same block. This tractor had been demonstrated to senior Massey-Harris executives and one of them, G.T.M. Bevan, European Division vice-president, commented in a letter dated 4 April 1953, that the LTX had a 'fantastic performance'. He added, 'This tractor will possibly be at the Smithfield Show'. One of the LTX prototypes was shipped to Detroit in about June 1953 for an evaluation programme for the American market.

Correspondence in the Massey-Ferguson archives shows that the three- to four-plough LTX had been a major attraction to the Massey-Harris

82

THIS IS THE
517651st TRACTOR
AND THE LAST
OF THE
TE 20 BUILD

The last of the TE tractors off the line at Coventry.

directors considering the possibility of purchasing Ferguson's companies. Their interest was doubtless sharpened when Harry Ferguson sent a letter to his British distributors on 13 May 1953 announcing the new 48 hp tractor and inviting orders for Spring 1954 delivery.

After the sale of his business, Harry Ferguson had assumed that this tractor would be rushed into production as the TE 60, and when this was delayed he started a campaign to bring the project forward. This situation eventually led to his resignation as chairman and the decision to sell his shareholding back to Massey-Harris.

During the next 5 years or so, Harry Ferguson and the engineering research company he had set up at Coventry concentrated on developments for the motor industry. These included four-wheel drive systems and a torque-converter transmission, which were both taken up by manufacturers. His work on four-wheel drive was many years ahead of the current interest in this idea as a means of improving traction and safety on the road.

Meanwhile, in 1958, Ferguson had the satisfaction of knowing that three of his TE 20 tractors had successfully completed a 1200-mile journey through some of the most difficult conditions in the world to reach the South Pole. The expedition was part of a British Commonwealth research programme, and was led by New Zealander Sir Edmund Hilary.

In the late 1950s Ferguson was again ready to revolutionise the tractor industry. In November 1959 Albert Thornbrough, Massey-Ferguson president, accepted an invitation to visit Ferguson's home, Abbotswood, to see a new prototype tractor at work. The main purpose of the demonstration was to show the benefits of a limited-slip differential incorporated into the drive system to improve traction in difficult conditions.

Thornbrough was impressed, both by Ferguson and by his tractor. Harry Ferguson, in his 76th year, was in 'good form', Thornbrough reported, but had some difficulty attaching implements to the tractor because his eyesight was failing.

This was the start of new negotiations between Ferguson and the Canadian company which almost ended with another manufacturing agreement. The new tractor would have a revolutionary but secret new transmission, later revealed to be a torque converter, would be manufactured in small, medium and large versions, and would have a flat-four engine design. Ferguson insisted that the production version of the tractor would be painted grey, and a draft agreement drawn up by Harry Ferguson Research stipulated that it would carry only the Ferguson name.

Correspondence continued from November to the following July, including offers of co-operation from Massey-Ferguson and a suggestion that the Canadian company should provide engineers to help with some of the development work.

That project, together with an automobile development he claimed to be negotiating in 1959 with Henry Ford II, never reached the market. Harry Ferguson's health was failing, and he died in October 1960, leaving behind a record of achievement which survives on farms throughout the world.

6 · AFTER THE MERGER

The task of merging the two companies was probably more difficult and more complicated than had been anticipated, and some mistakes were made. One of the mistakes appears to have been underestimating the urgency of integrating the management and engineering staffs in order to encourage a feeling of unity and the result was a long-lasting rivalry between the 'red' and the 'grey' factions in the company.

It was also important to make sure that the two tractor ranges and the dealer networks were brought together smoothly and effectively. This happened in Britain, where Massey-Harris machinery and the MH 744 and 745 tractors were obviously complementary to the smaller Ferguson tractors and equipment, although the grey and red colours were retained until November 1957.

In the United States the situation was more complex and the company appears to have aggravated matters by an attempt to continue the devel-

The TO tractor from the Detroit factory.

A TO 35 tractor equipped with an experimental radio-control system.

opment of the Ferguson and Massey-Harris product lines and dealer networks separately and in direct competition with each other. One result of this policy was a series of new tractor models to strengthen both product lines, and these helped accentuate the grey and red divisions.

The Ferguson TO 30 tractor, which had been in production at the Detroit factory since the end of the Ferguson-Ford agreement, needed updating and this project was given top priority. The new model was the TO 35, which arrived on the market in January 1955 with the familiar styling of the earlier Ferguson TE and TO models. Important improvements included a new gearbox with six forward ratios instead of the four available on the TO 30, and a Continental Z 134 engine of 133.6 in³ instead of the 129 in³ capacity of the TO 30 engine. The new model also included a live power-take-off among the equipment options.

The TO 35 retained the Ferguson name, styling and grey paint colour, but a comparable model was added to the production line for the Massey-Harris dealers to sell. This was the MH 50, which was given familiar Massey-Harris styling and carried the Massey-Harris name and red colour scheme.

Underneath the red paint, the MH 50 was remarkably similar to the TO 35. The gearbox provided six forward ratios and the engine capacity

Above right:
Ferguson 35 tractors were made in Coventry with a grey and gold paint finish.

Right:
Ferguson 35 from the Massey-Ferguson Museum.

86

was 133.6 in^3. The MH 50, which was launched almost 12 months after the new Ferguson model, was equipped with what the marketing experts called Hydramic Power. This, the sales leaflet explained 'combines the power developed in the engine, the pressure created by the hydraulic pump, and the tension and compression applied to the Draft Monitor 3-point Hitch in a system of automatic weight transfer'. This was a somewhat complicated formula to give the Massey-Harris tractor range the benefits of the Ferguson System without using the Ferguson name.

Another 1956 development for Massey-Harris dealers was a new lease of life for the old MH 33, 44 and 55 models. The new versions were called the MH 333, 444 and 555, and they were offered with the old Depth-O-Matic hydraulics and with power steering as an option. A significant improvement was the addition of a transfer box to double the number of gear ratios.

During the period before Harry Ferguson sold his companies to Massey-Harris, his engineers had been developing more equipment to sell with the Ferguson tractors. These included an interesting series of machines which were designed to be mounted on the TE and TO tractors and powered from the power-take-off. There was a forage harvester, a combine harvester and the Tractor-Mate baler.

Some of this development work continued after the Massey-Harris takeover, and the baler and the forage harvester were both added to the Ferguson product line. Sales were small, partly because of some mechanical problems and also because the output of both machines was restricted by the low power available from the small Ferguson tractors and was insufficient for most farmers.

The TO 35 tractor left Ferguson dealers in the United States without a

Rowcrop version of the MH 333 tractor.

The side-mounted Ferguson Tractor-Mate baler was available for the 1955 season.

Above left:
MH 444 tractor with rowcrop front wheels.

Left:
The MH 555 was the biggest tractor in the Massey-Harris range.

rowcrop-type tractor to sell against the new MH 50 and, to remedy this, the Ferguson F40 was rushed into the product line. It was the first tractor sold under the Ferguson name in both rowcrop and tricycle versions and with clearance for mid-mounted equipment.

The Ferguson F40 was given a three-plough rating, but used the same 133.6 in³ engine as the TO 35. Production at the Ferguson factory in Detroit started in 1956 and finished in the following year when less than 10 000 had been built.

This was the last model designed specifically to continue the policy of selling two tractor ranges. In 1957 the MH 50 was replaced by the MF 50 and, later in the same year, the Ferguson 35, the British version of the TO 35, changed its grey and gold paint for Massey-Ferguson red and changed its model number to MF 35. At the 1957 Smithfield Show in London an additional model was announced from the Coventry factory. This was the MF 65, powered by a Perkins diesel rated at 50.5 hp.

Special features of the MF 65 included self-energising disc brakes which

Left, below and right:
The Ferguson TO 40 made a brief appearance in standard, tricycle and rowcrop versions.

Right:
The Massey-Harris name was still used for the MH 50 tractor.

were mounted inboard, a live power-take-off, and power steering available as an option. The gearbox for the MF 65 was the same as that on the MF 35 but operating at higher speeds to cope with the extra horsepower.

In 1958 the Massey-Harris-Ferguson company name which had been agreed during the negotiations with Harry Ferguson, was abbreviated to Massey-Ferguson.

At that time the MF 35 and 65 models gave the company a strong position in what was then the small-to-medium sector of the market, and this was later strengthened again by the addition of the diesel-powered 25 hp MF 25 model from the Beauvais factory in France.

The situation was much less satisfactory further up the market, and the Massey-Ferguson tractor range was ill-equipped to cope with competition from American-based manufacturers with tractors of 60 hp and above. Another new model, the MF 85, with a 60 hp rating and the choice of petrol or diesel engines, arrived in 1959 to compete further up the power scale but, although this model proved to be popular, it was too small to fill the power gap at the top of the market.

This presented a serious problem for the company. The demand for tractors of more than 60 hp was increasing fast, particularly in the all-important North American market, and this was a sector in which Massey-Ferguson had nothing to offer. The short-term solution to the problem was to buy in a suitable tractor from a rival manufacturer and

Right and below right:
The MF 25 for the North American market and the MF 825 from the Massey-Ferguson factory in France.

The MF 35 was developed from the Ferguson 35.

the first result was the MF 95 tractor which was supplied by the Minneapolis-Moline company. This tractor was powered by a 425 in³ engine with the unusual arrangement of six cylinders in three blocks of two cylinders each. The power rating was 65 hp on the belt at 1300 rpm and 63.5 hp at the drawbar at 1500 rpm. The first batch of MF 95 tractors was sold in 1958 and sales increased to about 700 a year before a replacement model, the MF 97, was introduced in 1961.

The MF 97 was a more powerful version of the 95 and was available with a four-wheel drive option which, the sales leaflet claimed, added '30 to 50 per cent more traction'. Sales of the MF 97 tractor peaked at 1756 in 1963 before this model was withdrawn in 1965. The MF 97 was the first tractor sold by Massey-Ferguson with 100 hp on the belt.

Massey-Ferguson also bought in a batch of 500 tractors from the Oliver Corporation, and these were sold as the MF 98 in 1959 and 1960. This model arrived on the market before the MF 97, and was announced by Massey-Ferguson as the most powerful tractor they had ever offered. It was equipped with an unconventional power unit with three cylinders of 213 in³ capacity. It was built by General Motors and was described in the Massey-Ferguson sales leaflet as a blown two-stroke diesel. The leaflet claimed that two-stroke diesels were more efficient and also more reliable than four-strokes – which appears to suggest that the four-stroke engines used in other Massey-Ferguson tractors were second best!

MF Super 90 tractors were announced in August 1962.

Another of the rather confusing 90 series models brought into the Massey-Ferguson model range to compete at the top end of the market was the MF Super 90, a tractor built by Massey-Ferguson and available with a 242 in³ petrol engine or a 302 in³ Perkins diesel. The tractor was in production from 1962 to 1965 with a maximum power output of 68 hp.

While the big 90 series tractors were giving Massey-Ferguson a presence among the companies supplying big tractors, the company was making long-term plans for a completely new tractor range.

The immensely successful MF 135.

This was the DX project, an ambitious development programme which started in 1962. Professor Neufeld, in his book *A Global Corporation*, said that the Engineering Division allocated 1 million man-hours to design work, prototype building and the test programme for the new tractors. The project included six different models, all of which were either completely new or were substantially redesigned versions of earlier models.

The aim of the DX programme was to give Massey-Ferguson, for the first time, a model range which was technologically advanced, covered all the major sectors of the market and was designed for marketing on a completely international scale.

When the new range was announced in 1965 it included the MF 135 and MF 150 models with a maximum output of about 37 hp, the MF 165 with a 50 hp output, the MF 175 providing about 60 hp, plus the big MF 1100 and 1130 models with more than 90 hp and a maximum of 120 hp respectively. The big tractors were built in North America and the small and medium models were sourced from British and French factories.

Further models were added to the tractor range produced under the DX programme, mainly to keep pace with the market demand for extra power. An example was the 80 hp 1080 tractor, and the company also achieved considerable sales success with a series of big tractors with four-wheel drive through equal-sized wheels and articulated steering. The biggest of these were the American-built MF 1505 and 1805 tractors, powered by Caterpillar V 8 engines producing up to 175 and 192 hp respectively. These models were in production from 1975 and were eventually replaced by the 4000 series, the first tractors with eletronic linkage control.

The most popular of the articulated four-wheel drive models in this series was the MF 1200 and its later 1250 version. The 110 hp engine put the MF 1200 in a popular power range for big farms, and this was the model which helped to establish the American four-wheel-drive system with equal-sized wheels in Europe.

7 · PERKINS, STANDARD AND LANDINI

The sale of the Ferguson tractor interests to Massey-Harris was one of the most important commercial development in the history of the tractor industry, and it was followed by a series of additional purchases during the next 10 years which consolidated and extended the Massey-Ferguson company's tractor business.

It was during this period that F. Perkins Ltd of Peterborough, one of the leading companies in diesel engine technology, became part of the Massey-Ferguson group.

Frank Perkins had started the company in 1932 with Charles Chapman as his chief engineer. Both had previously been employed at Aveling and Porter of Rochester, Kent, a large engineering company which began development work on a four-cylinder diesel engine in the late 1920s. Charles Chapman and Frank Perkins were closely involved in the diesel project which produced the Invicta engine, used in small numbers to power tractors built by the Garrett company of Leiston, Suffolk, and for trucks and buses.

The Aveling and Porter company and Garrett were both members of a big consortium of engineering companies called Agricultural and General Engineers, formed to provide a British group with sufficient resources to compete against the big American farm equipment manufacturers. By 1932, Agricultural and General Engineers was bankrupt, taking a large section of the British agricultural engineering industry with it. Companies within the group were sold and some, including Aveling and Porter, eventually prospered again, but the damage caused by the collapse was immense and the Invicta-powered diesel tractor was one of the casualties.

The Invicta diesel, for which Charles Chapman was responsible as head of engineering at Aveling and Porter, proved to be an excellent design and, in 1930, it powered a Garrett tractor in a remarkable demonstration of reliability.

This was a non-stop ploughing marathon which was carried out under the supervision of staff from the Institute of Agricultural Engineering attached to Oxford University. The previous non-stop ploughing record was 481 hours, and the Garrett tractor beat this easily by working 977 hours without a break. At that point the engine was stopped by accident, but it was restarted immediately and continued for a further 600 hours.

In spite of this demonstration of reliability, plus an impressive performance at the World Tractor Trials held near Oxford in the same year, few of the diesel-powered Garrett tractors were built as the AGE group lurched towards financial collapse. When the end came in 1932, Frank

The famous Perkins P6 diesel engine

Perkins decided to start his own company to continue the development of diesel engines and he moved to Peterborough with Charles Chapman to start work on a small engine suitable for vans and light trucks.

By the end of 1932 the first Perkins engine, called the Vixen, was being tested. It was a four-cylinder design which was improved and uprated to become the Fox and, later, the Wolf series. There was also a larger engine which was called the Leopard, with 100 mm cylinder bore compared with the 75 mm of the Vixen and 110 mm of the 42 hp Aveling and Porter engine.

Although the pace of technical development appears to have been rapid, commercial progress was slow and it was 5 years before the company started to make a profit. At this stage most of the engines were being fitted to commercial vehicles as an alternative to the standard petrol units. There was also a growing market for marine versions of the engines, but the market for multi-cylinder diesels for farm tractors in Britain had not developed, in spite of the excellent results achieved by this type of engine at the 1930 World Trials.

In 1937 the company completed the development of its first six-cylinder diesel, the P6, an engine which later became immensely successful and set new standards for efficiency and reliability. During World War 2 the company built diesel engines for the armed forces, including a contract to supply the power units for naval rescue craft but, soon after the war had ended, the demand for diesel tractors started to develop and the P6 became the most popular diesel engine for tractors in Europe. Customers included Ford, who offered the P6 as a more powerful alternative to their own petrol-paraffin engine in the Fordson E27N, and Massey-Harris, who used the P6 as the power unit for the MH 744 tractor, the British-built version of the MH 44.

Massey-Harris chose the P6 after a thorough investigation of alternative engines available from sixteen different manufacturers. This evaluation took place during October and November 1947 and included several visits to the Perkins factory and demonstrations of Fordson and Massey-Harris tractors powered by the P6.

The Massey-Harris company president, James Duncan, was one of the visitors to the Perkins factory and the notes he made later show that he

A modern Perkins six-cylinder engine with turbocharger.

was impressed by what he saw. He particularly approved of Frank Perkins, he was impressed by the efficiency in the factory, and he noted that the capacity to build 12 000 engines a year was failing to keep pace with demand and the company was seeking a Government permit to expand production to 25 000 a year.

Massey-Harris continued to specify Perkins engines throughout the production life of the MH 744 tractor and remained a customer after the Ferguson takeover.

The first major setback in the Perkins success story came in the 1950s partly because the launch of a new engine caused problems and partly because of increased competition from other engine manufacturers. The problems were sufficiently serious to persuade the directors to look for some form of association with another company. Discussions during 1958 resulted in an agreed offer from Massey-Ferguson to take over the Perkins company, and this was announced in January 1959.

This arrangement was a logical move for both companies. It gave Massey-Ferguson complete control over its source of engines and it provided Perkins with the finance needed to develop new engines and, later, to extend the factory to allow increased production. An important objective at the time of the takeover was for Perkins to continue to supply engines to other customers, including tractor and combine-harvester manufacturers in direct competition with Massey-Ferguson, and this has remained a basic part of the company policy.

One of the prototype Standard tractors.

With the purchase arrangements for the Perkins engine business completed, the next major development in the Massey-Ferguson tractor operations followed quickly. This was an agreement to take over the production facilities which the Standard Motor Company had established in England and France to make tractors for Harry Ferguson.

The agreement between Harry Ferguson and the Standard company had worked reasonably well. The Ferguson tractor provided Standard with a product which they could manufacture in large volume and which fitted in well with their car business. It was also a product which appears to have contributed substantially to their profits and this seemed likely to continue as Ferguson showed little interest in setting up his own factory in England to make tractors.

The situation appears to have deteriorated after Massey-Harris arrived on the scene. One of the problems for the Standard company manage-

Rear linkage on a Standard tractor.

ment was that the car business was moving into an increasingly competitive situation and they needed finance to keep their Standard and Triumph car ranges up to date and to tool up for new tractor models.

There were frequent disagreements between the two sides over production costs in the Banner Lane tractor plant and this helped to encourage friction and lack of confidence. It also encouraged the idea that Massey-Ferguson should take over the tractor manufacturing business. At about this time the Standard management realised they might lose the tractor business, and they decided to develop a small Standard tractor which they would be able to put into production if they were forced to find a replacement for the Ferguson. The tractors were designed around the Standard diesel engine and one of the prototypes has survived and, at the time of writing, is in the Marshall Tractors collection.

A takeover was discussed several times and Massey-Ferguson bought enough stock to become the biggest shareholder in Standard but it took several years of negotiation before the basis for a takeover agreement was completed in July 1959. Under the terms which were eventually agreed, Massey-Ferguson purchased the Standard company's tractor assets in England and France for a cash payment of about £32 million plus the $8 million or so of Standard shares which had been bought by Massey-Ferguson. The main asset acquired in the transaction was the big Banner Lane factory at Coventry with the capacity to build 100 000 tractors a year. There was also a French factory at Beauvais, but the Standard engine facility was not included because Massey-Ferguson now owned the Perkins factory.

A third major development during this period of rapid expansion came in 1960 when Massey-Ferguson bought its way into the Italian tractor industry by taking over the Landini company, with a tractor factory at Fabbrico and engine production facilities near Como.

The company was started by Giovanni Landini, who established a blacksmith's business in Fabbrico in 1884, just 7 years before the Massey and Harris companies had joined forces in Canada. He started doing repair work for local farms, but soon progressed to making simple equipment which was winning medals at local shows and exhibitions from 1888 onwards.

By 1910 Landini had designed and built his first engine, a single-cylinder semi-diesel which was designed to operate stationary equipment and, in the following year, he built a portable steam engine suitable for driving a threshing drum.

Engines became an increasingly important part of the Landini company product range and, in the early 1920s, Giovanni Landini and his sons became interested in the idea of designing a tractor. Giovanni died in 1924 and his sons took over the business and the tractor project.

The first Landini tractor was completed in 1925. The power unit was a single-cylinder engine operating on the semi-diesel principle which was popular among German, Italian and Swedish manufacturers, but almost unknown in American and British tractors.

Giovanni Landini.

Semi-diesel engines used in tractors were almost always single cylinder and, as in the Landini tractors, they were usually designed to operate in the horizontal position, with the cylinder head toward the front end of the tractor. The biggest attraction of this type of engine was its extreme simplicity and this made it relatively cheap to manufacture. It was also a basically reliable design, operating at slow speeds and with few working parts to cause problems.

Another attraction during the 1920s and 1930s was the fact that semi-diesels operate on almost any type of liquid fuel. This meant that owners could use whatever was cheap at the time; this might include a mixture of paraffin and waste sump oil, for example.

These attractions helped to compensate for the fact that semi-diesels are notoriously dirty and lack the typical smoothness of a multi-cylinder diesel or spark-ignition engine. They also have a low operating efficiency with a large cylinder capacity to produce a modest power output, and the early Landini tractor engines provide some good examples of this.

The 40 hp tractor announced by Landini in 1932 was powered by a semi-diesel engine with a massive single cylinder of 14,300 cc capacity. This is an output of just 2.8 hp per litre, and compares with the 5.3 hp per litre developed by the four-cylinder petrol-paraffin engine used in the Wallis-based 20-30 tractor which Massey-Harris was selling at that time. The Aveling and Porter diesel engine, forerunner of the early Perkins diesels, developed 6.69 hp per litre when it was tested at the World Trials

The 40 hp Landini tractor
announced in 1932.

in 1930. A modern Perkins diesel, such as the TV8.540 turbocharged V-8, develops up to 235 hp from 8.83 litres, or 26.6 hp per litre.

Maximum power from the 40 hp engine was developed at a leisurely 500 rpm. Like all Landini tractors built before 1941 the gearbox of the 40 hp model provided three forward speeds and a reverse. The tractor weighed 2.8 tonnes.

The best known of the early Landini tractors were the 48 hp Super Landini of 1934, and the Velite which was in production from 1935. These were the models which established Landini as one of the leading companies in the European tractor industry during the 1930s, and both are regarded as classic examples of pre war design.

Super-Landini production started with seventy-nine tractors completed in 1934 and continued until 1951 when the production total had reached 3530. Each tractor weighed 3.6 tonnes and the maximum engine speed of 620 rpm provided a top travel speed of 3.9 mph in third gear. The engine was a typical single cylinder, semi-diesel but with the output improved to 48 hp from 12.2 litres or 3.9 hp per litre.

When the Super Landini arrived on the market it was Italy's biggest tractor and one of the most powerful tractors in Europe. The Velite was smaller, weighing 2.4 tonnes with a full tank and with only 7 222 cc of engine capacity to produce the 30 hp maximum output. It was also a more spritely performer than the Super model with 5.3 mph top speed when the engine was running at its 720 rpm maximum.

An unusual feature of the Super Landini and the Velite is that both models were supplied with electric lights front and rear, equipment which few manufacturers could offer in the 1930s.

Landini continued to rely on the semi-diesel engine long after most tractor companies had switched to other types of power unit. There were still customers in Italy and in some of the company's export markets who

A refill for the tank of this Super Landini.

In 1941, Landini announced the
Butalo tractor.

preferred the economy, reliability and easy servicing of the semi-diesel,
and tractors powered by this type of engine were still in production at the
Fabbrico factory in 1961.

The decision to bring another type of engine on to the production line
to offer as an alternative to the semi-diesel was taken in 1955. By this time
most of the leading manufacturers had at least some diesel-powered
models available and this was the obvious engine choice for Landini.
Instead of spending time and large amounts of money on developing a
range of diesel engines, the company arranged a licence agreement allow-
ing Landini to manufacture some of the Perkins diesels in Italy. Production started at the Como factory during 1956 and, for the next 5 years, the
demand for Landini diesel tractors increased while sales of models
powered by the old semi-diesel engines dwindled rapidly to nothing.

With the diesel engines in production and helping to improve the
company's sales performance, the decision was taken to add crawler tractors to the product range for the first time. Crawler tractors are popular in
Italy, partly because there are large areas of steep land, where the extra
stability of a tracklayer is valued, and partly because the tradition of deep
ploughing on Italian farms makes good use of the excellent traction
available from crawler tracks.

The first Landini crawler model was the C25 which was announced at
the 1957 Verona Show, although it took another 3 years of development
before the tractor was ready for the market. The C25 was powered by a

25 hp diesel engine and used the new up-to-date styling which was brought into the Landini range when the diesel models arrived.

One unexpected result of the agreement with Perkins was that it brought the Landini company into contact with Massey-Ferguson. This happened in 1960, less than 12 months after Perkins had become part of the Massey-Ferguson group and negotiations were started almost immediately for Massey-Ferguson to take over the Italian company. Landini became part of the Massey-Ferguson group during 1960.

For Landini, the takeover brought long-term security at a time when some of the smaller independent tractor companies were starting to face increasing competitive pressure from the big international manufacturers. In Landini's case there was additional concern that the import tariffs which protected Italian manufacturers were being lowered because of Italy's membership of the European Economic Community.

For Massey-Ferguson the investment in Landini offered substantial benefits. Italy had been a difficult export market, mainly because of the tariff barriers, and Landini offered Massey-Ferguson a way into the market with a respected name and an established dealer network. Landini was also well established in some export markets which were of interest to Massey-Ferguson, and there was the additional advantage that the new Landini tractors had been designed around Perkins engines.

Another of the attractions which persuaded Massey-Ferguson to invest in Landini was the new C25 and other crawler tractors under development at Fabbrico. This was still an important sector of the agricultural tractor market and the Landini deal offered the first opportunity to sell crawler tractors.

An important factor in the purchase arrangements was that the Italian company would continue to use the Landini name, which was an important asset, and the tractors would continue to look distinctive. The Landini paint finish is a bright blue, but red is used for some tractors which will be sold overseas under the Massey-Ferguson name.

In 1984 the Landini company celebrated its centenary and the following year was its sixtieth as a tractor manufacturer.

Varity Corporation sold a controlling interest in the Landini company to a Belgian based finance and industrial group in January 1990. Under the new arrangement Landini is continuing to supply tracklaying, orchard and vineyard tractors in MF colours, and Landini tractors continue to have engines supplied by Perkins.

8 · A GLOBAL CORPORATION

It took Massey-Harris three attempts over more than a decade to make a modestly successful start in the tractor industry. Now, more than half a century later, the company has become the western world's largest tractor manufacturer.

Much of this success was brought in as other companies were purchased to contribute to the development and expansion of the tractor operations. This process began with the Wallis tractors with their unique U-frame design, took a big step forward when the Ferguson System became part of Massey-Harris, and was followed by further expansion with the purchase of Landini and the addition of the Standard factories in Britain and France.

The result is a business which operates on a global scale with about 9000 distributors and dealers, with tractor factories in Britain, France and Italy,

The MF 230 tractor achieved outstanding fuel economy at Nebraska.

and with a product range from about 15 to 320 hp including compact models bought in from a Japanese supplier. The chief executive is Victor Rice, the first British born person to hold this position.

Tractors sold under the Massey-Ferguson and Landini names claimed more than 17 per cent of the available world market in 1985, significantly more than any other tractor company. The tractor operations include some of the most advanced facilities available, with a recent investment of more than £20 million to extend the use of computer-aided design equipment and increased automation at the Banner Lane factory at Coventry.

Banner Lane builds almost 10 per cent of the tractors sold in the West and almost half of all the tractors manufactured in Britain, with the grand total approaching the 3 million mark since 1946 when the first of the Ferguson TE series came off the production line. The tractors produced at Banner Lane are the big volume models up to 95 hp, with the higher horsepower 3000 and 2005 series made at the French factory in Beauvais. The Landini factory at Fabbrico makes mid-range wheeled tractors from about 50 to 85 hp, and is also responsible for manufacturing 'specials' to be sold under both the MF and Landini names, including all the crawler models, orchard and vineyard tractors.

The acquisition policy, which helped to put Massey-Ferguson at the

MF tracklayer built at the Landini factory.

top of the world tractor league, is still contributing to the company's development. In 1985 the combine harvester interests of the White company of Canada were taken over to contribute advanced rotary separation technology to the top end of the Massey-Ferguson combine range and, in 1983, the Rolls Royce diesel engine factory at Shrewsbury was purchased to add to the range available from Perkins.

The combination of the Perkins and Rolls Royce engine lines has produced one of the biggest and most comprehensive series of diesel engines in the world. Rolls Royce has concentrated on the medium to large sector of the market, with engines of six to twelve cylinders developing from about 200 to 1200 hp. The main markets for the big diesels are trucks, industrial equipment such as generating sets and railway locomotives, and military vehicles, including the Challenger battle tank.

Engines from Peterborough cover the market from about 3.5 to 300 hp and are used for a wide range of agricultural, industrial, marine and vehicle applications. Perkins engines are used by more than 500 equipment manufacturers in 160 countries.

A major development in 1986 was the announcement of the Perkins Prima engine, claimed to be the first small high-speed diesel with direct injection. Prima was designed to power cars and light commercial vehi-

One of the giant tractors from the MF factory in Canada.

MF 699 tractor equipped with front
and rear spray equipment and
Terra-Tires.

Right:
The atmospheric test chamber at
Peterborough can simulate any
climatic conditions from Arctic cold
to tropical heat.

Perkins test facilities include this
special chamber for noise measurement.

cles and is the result of a joint venture between Perkins and the Austin Rover Group.

Direct injection allows the fuel-and-air mixture to go directly into the combustion space in the top of the piston. This is more efficient than the indirect-injection system where combustion occurs in a pre-chamber in the cylinder head. Direct injection is widely used in tractor engines, for example, but the Perkins Prima is the first direct-injection diesel developed to operate at around 4500 rpm, the speed required from diesel engines used in cars. Prima is a joint venture between Perkins and Austin-Rover, and has been claimed to be the most advanced engine of its kind available anywhere.

Other major developments during 1986, one of the most important years in the company's history, were the launch of the new 300 range and the 3000 series with their advanced electronics, and the announcement on

The MF 698T was the first tractor with a 'wastegate' turbocharger.

One of the production lines at the Massey-Ferguson factory in Coventry.

14 May that Massey-Ferguson Ltd was changing its name to Varity Corporation.

The decision to change the name was taken partly to signify the company's recovery from a period of serious financial problems and into a new period of development, and also because the Massey-Ferguson name was considered to be too closely associated with the farm equipment business for a company which was developing a wider spread of major commercial interests.

The Varity name has no special significance, apparently, although it is no coincidence that it has some similarity to Verity, the name of a Canadian implement company in which Massey-Harris acquired a major shareholding in 1892.

Within Varity Corporation the Massey-Ferguson name is retained for the tractor and farm machinery division. Other divisions include Perkins Engines, Massey-Ferguson Industrial Machinery and Pacoma Components which manufactures hydraulic items such as valves and cylinders. Varity Enterprises is the part of the group which is responsible for finding new investment opportunities.

Although the Varity name is part of a long-term plan to develop and strengthen the group, the tractors continue to carry the Massey and Ferguson names — two of the men who have made important contributions to the development of farm mechanisation throughout the world.

APPENDIX 1
FERGUSON SYSTEM PATENTS

Application Date	Patent Number	Feature Covered
12 Sep. 1917	119883	Arrangement for levelling ploughs using a system of cranks to support the plough.
13 Dec. 1919	160248	Implement depth control through a linkage to the tractor, using parallel links to achieve weight transfer.
28 Dec. 1921	186172	A development of the previous patent, but incorporating a depth wheel and with links not necessarily parallel.
3 Nov. 1921	195421	Also based on 160248, but with spring-assisted mechanism to maintain depth.
11 Dec. 1923	226033	An arrangement to avoid carrying the full weight of an implement on the depth wheel.
12 Feb. 1925	253566	Provision for either an electric motor, friction clutches or fluid pressure for the lift and depth control mechanism.
3 Jul. 1928	320084	Converging linkage geometry with two top link arms and one lower arm.
2 Feb. 1937	470069	Hydraulic control valve on the suction side of the pump to avoid aeration and overheating in neutral or zero-flow position.
5 Feb. 1937	470087	Automatic release of hydraulic pressure for protection when an implement hits an obstruction.
30 Nov. 1937	510352	Power-take-off shaft located within the triangle formed by the linkage geometry.

APPENDIX 2
PRODUCTION DATA

In the following pages are given the production years, models and serial numbers of tractors manufactured by Wallis, Case, Massey-Harris and Ferguson.

Wallis Tractors (J.I. Case Plow Works)

1912–1913	Bear	201–210
1913–1917	Cub C & D	1001–1660
1915–	Cub J	10001–13505
1916–	Cub K	14001–23156
1922–	Cub OKO	23200–25644
1926–	Cub Cert. Std.	25645–40000
1926–	Cub OKO	40001–50000

Massey-Harris

1929	12	100101–107000
1930	12 OPA	200000–200250
1930	12 Ind	250001–250026
1930–1936	Gen. Purp.	300000–303000
1931–1938	25	69001–73111
1937	25 Ind	90001–90200
1938	25	85001–90200
1936	Gen. Purp. (new)	303001–355000
1936–1937	Pacemaker	107001–109837
1936	Pacemaker OPA	200251–200402
1936	Challenger	130001–133366
1938	Pacemaker	109838–130000

1938	101 Sr	255001–256084	355001–355602
1939	101 Sr	256085–257280	355603–356791
1940	101 Sr	257281–258768	356792–358187
1941	101 Sr	258769–259761	358188–358868
1942	101 Sr	259762–260429	358869–358974
1943	101 Sr	260430–260795	358975–359457
1944	101 Sr	260796–263019	259458–360926

1945	101 Sr	263020–264999	360927–362519	
1946	101 Sr	270145–272506	362520–363647	
1939	101 Jr	375001–376157		377001–377927
1940	101 Jr	376158–377000	395001–395569	377928–379549
1941	101 Jr	395570–397636		379550–379814
1942	101 Jr	397637–398595		379815–379854
1943	101 Jr	398596–399999	500001–500002	379855–380640
1944	101 Jr	500003–502433		380641–382568
1945	101 Jr	502434–503778		382569–384297
1946	101 Jr	503779–505513		384298–385641
1941	102 Sr	365001–365201		
1942	102 Sr	365202–366061	265001–265043	
1943	102 Sr	366062–366182	265044–265077	
1944	102 Sr	366183–367352	265078–265287	
1945	102 Sr	367353–367423		
1940	201	91201–91703	98001–100131	
1941	202	95001–95223		
1939	102 Jr	385001–385203		387001–387030
1940	102 Jr	385204–385449		387031–387126
1941	102 Jr	385450–386098		387127–387418
1942	102 Jr	386099–386661		387419–387600
1943	102 Jr	386662–387000	390001–390007	387601–387843
1944	102 Jr	390008–390993		387844–388239
1945	102 Jr	390994–391912		388240–388994
1946	102 Jr	391913–392749		388995–389172
1940	203	91201–91541	95000–95001	98001–98027
1941	203	91541–91690	95002–95181	98028–98363
1942	203	91691–91702	95182–95222	98364–98673
1943	203			98674–98806
1944	203	95224–95342	95223–95294	98807–99688
1946	203		95295–95337	99689–100119
1947	203		95338–95342	100120–100131
1941	81	400001–403167	425001–425677	
1942	81	403168–403353	425678–425779	
1944	81	403354–403363	425757–425779	
1945	81	403364–404663	425780–426802	
1946	81		426803–427574	
1948	81	404664–406601		
1941	82	420001–420054	435001–435278	

1942	82		420055–420273	435279–435451		
1943	82			435452–435457		
1945	82		420274–420306	435458–435737		
1946	82		420307–420425	435738–436159		

1946	20 gas r/c	1001–1579	20 gas std	1001–
1947	20 gas r/c	1580–3583	20 gas std	1002–2229
1948	20 gas r/c	3584–5195	20 gas std	2230–2661

1947	20 k r/c	1001–1353	20 k std	1001–1818
1948	20 k r/c	1354–1644	20 k std	1819–2431

1948	22 gas r/c	1001–2095	22 gas std	1001–1541
1949	22 gas r/c	2096–4579	22 gas std	1542–3207
1950	22 gas r/c	4580–7623	22 gas std	3208–4532
1951	22 gas r/c	7624–10136	22 gas std	4533–5716
1952	22 gas r/c	10137–10783	22 gas std	5717–6253

1948	22 k r/c	1001–1153	22 k std	1001–1316
1949	22 k r/c	1154–1335	22 k std	1317–1487
1950	22 k r/c	1336–1557	22 k std	1488–1569
1951	22 k r/c	1558–1775	22 k std	1570–1747
1952	22 k r/c	1776–1827	22 k std	1748–1823

1946	30 gas r/c	1001–	30 gas std	1001–
1947	30 gas r/c	1002–3385	30 gas std	1002–2119
1948	30 gas r/c	3386–6824	30 gas std	2120–3193
1949	30 gas r/c	6825–9344	30 gas std	3194–5566
1950	30 gas r/c	9345–13815	30 gas std	5567–7490
1951	30 gas r/c	13816–17933	30 gas std	7491–8695
1952	30 gas r/c	17934–19382	30 gas std	8696–9337

1947	30 k r/c	1001–1224	30 k std	1001–1893
1948	30 k r/c	1225–2009	30 k std	1894–3250
1949	30 k r/c	2010–2392	30 k std	3251–3530
1950	30 k r/c	2393–2718	30 k std	3531–3860
1951	30 k r/c	2719–3179	30 k std	3861–4170

1946	44 gas r/c	1001–	44 gas std	1001–1140
1947	44 gas r/c	1002–2047	44 gas std	1141–1870
1948	44 gas r/c	2048–5317	44 gas std	1871–4527
1949	44 gas r/c	5318–13821	44 gas std	4528–9580
1950	44 gas r/c	13822–21814	44 gas std	9581–13725
1951	44 gas r/c	21815–31189	44 gas std	13726–17058
1952	44 gas r/c	31190–33890	44 gas std	17059–19117

1946	44 k r/c	———————	44 k std	1001–1010

Year	Type		Range	Type		Range
1947	44 k	r/c	1001–1078	44 k	std	1011–1440
1948	44 k	r/c	1079–1855	44 k	std	1441–3597
1949	44 k	r/c	1856–2598	44 k	std	3598–4826
1950	44 k	r/c	2599–3328	44 k	std	4827–6018
1951	44 k	r/c	3329–4203	44 k	std	6019–6785
1952	44 k	r/c	———	44 k	std	6786–6822
1948	44 dsl	r/c	———	44 dsl	std	1001–1022
1949	44 dsl	r/c	1001–1003	44 dsl	std	1023–2179
1950	44 dsl	r/c	1004–2482	44 dsl	std	2180–3988
1951	44 dsl	r/c	2483–4703	44 dsl	std	3989–5638
1952	44 dsl	r/c	4704–5656	44 dsl	std	5639–6396
1952	44 but.	r/c	1001–1460	44 but.	std	1001–1035
1950	44 GS vin.		1001–1030	44 GS orch.		1001–1100
1951	44 GS vin.		———	44 GS orch.		1101–1120
1950	44 DS orch.		1001–			
1951	44 DS orch.		1002–1034			
1951	44 GRA hi alt.		1001–1163	44 GSA hi alt.		1001–1054
1952	44 GRA hi alt.		1164–1269	44 GSA hi alt.		1055–1065
1946	44-6 gas	r/c	1001–	44 6 gas std		———
1947	44-6 gas	r/c	1002–2982	44-6 gas std		1001–2000
1948	44-6 gas	r/c	2983–4754	44-6 gas std		2001–2600
1949	44-6 gas	r/c	4755–5254	44-6 gas std		2601–2729
1950	44-6 gas	r/c	5255–5508			
1951	44-6 gas	r/c	5509–5928			
1952	44 series		40001–43699	$44 series		———
1953	44 series		43700–47060	$44 series		50001–51363
1954	44 series		———	$44 series		51364–58066

Year	Type	Range	Type	Range	Type	Range
1946	55 k	1001–1012	55 gas	1001–1115	55 dsl	———
1947	55 k	1013–1553	55 gas	1116–2131	55 dsl	———
1948	55 k	1554–3032	55 gas	2132–3580	55 dsl	———
1949	55 k	3033–4077	55 gas	3581–5467	55 dsl	1001–1021
1950	55 k	4078–4807	55 gas	5468–6398	55 dsl	1022–2057
1951	55 k	4808–5502	55 gas	6399–7077	55 dsl	2058–2821
1952	55 k	5503–5504	55 gas	———	55 dsl	2822–2964

Year	Type	Range	Type	Range
1949	55 GS rice & hill	1001–1034	55 KS rice & hill	1001–1012
1950	55 GS rice & hill	1035–1215	55 KS rice & hill	1013–1109
1950	55 D rice & hill	1001–1151	55 KS rice & hill	———
1951	55 D rice & hill	1152–1451	55 KS rice & hill	1110–1230

1952	55 D rice & hill	1452–1516	55 KS rice & hill	—————

1950	55 GS Western	1001–	55 DS Western	—————
1951	55 GS Western	1002–1082	55 DS Western	1001–1188
1952	55 GS Western	1083–1109	55 DS Western	1189–1212′

1951	55 GSA hi alt.	1001–1025	GSHA hi alt.	1001–1004
1951	55 GSWA hi alt.	1001–1004		
1952	55 GSWA hi alt.	1025–1026		

1952	55 series	10001–13016	1953 55 series	13017–15298
1954	55 series	15299–17058	1955 55 series	17059–17888

1952	21	1001–1416	22	20001–20584
1953	21	1417	22	20585–20623

1952	23	1001–1665	1953 23	1666–4345
1954	23	4346–4552	1955 23	4553–4772
1956	23	4773–4832		

1952	30	30001–30595	33	1001–2054
1953	30	30596–30600	33	2055–6616
1954	30	—————	33	6617–9781
1955	30	—————	33	9782–12607

1953	I-162 series	1001–1026		

Massey-Harris UK Production

1948	744	201–400	1949 744	401–1400
1950	744	1401–4400		

(After 1950 the production year is indicated by a letter, starting with F for 1951 and continuing with G for 1952, H for 1953 and so on.)

Ferguson Tractor Production

	Coventry		*Detroit*	
1946	TE series	1–316		
1947	TE series	317–20894		
1948	TE series	20895–77772	TO series	1–1800
1949	TE series	77773–116461	TO series	1801–14659
1950	TE series	116462–167836	TO series	14660–39162
1951	TE series	167837–241335	TO series	39163–72679
1952	TE series	241336–310779	TO series	72680–108644
1953	TE series	310780–367998	TO series	108645–125958
1954	TE series	367999–428092	TO series	125959–140001

FURTHER READING

The history of the Massey-Harris company during its first 100 years is described in Merrill Denison's *Harvest Triumphant*, published in 1948 by McLelland and Stewart. A more detailed corporate history of the company up to 1966 is available in Prof. E.P. Neufeld's *A Global Corporation* which was published by the University of Toronto Press in 1969.

Harry Ferguson – Inventor and Pioneer is an excellent book about the man who developed the Ferguson System. The book was published in 1972 by John Murray and includes considerable information about the inventor and the tractors which carried his name. For a more concise but well-illustrated account I recommend *Harry Ferguson*, a booklet by Bill Martin which is published by the Ulster Folk and Transport Museum.

Information about Landini history is available in two volumes published in 1984 to celebrate the centenary of the company. These are *Landini Cento Anni* by Andrea Bedosti and Giorgio Maioli, and *dal 1884 Landini*. Both volumes were published by MF-Landini and are superbly produced. The pictures are excellent, even if you don't understand the Italian text.

My own book, *Ford & Fordson Tractors*, published by Farming Press, includes additional information about the Ferguson System tractors built by Ford.

INDEX

Page numbers in *italic* refer to black and white photographs. Numbers in **bold** refer to colour plates.